Collins

SECURE SCIENCE FOR GCSE

Teacher Resource

INTERVENTION TO GET BACK ON TRACK

Lucy Wood

William Collins' dream of knowledge for all began with the publication of his first book in 1819. A self-educated mill worker, he not only enriched millions of lives, but also founded a flourishing publishing house. Today, staying true to this spirit, Collins books are packed with inspiration, innovation and practical expertise. They place you at the centre of a world of possibility and give you exactly what you need to explore it.

Collins. Freedom to teach.

Published by Collins
An imprint of HarperCollins*Publishers*
The News Building, 1 London Bridge Street, London, SE1 9GF, UK

HarperCollins Publishers
1st Floor, Watermarque Building, Ringsend Road, Dublin 4, Ireland

Browse the complete Collins catalogue at
www.collins.co.uk

10 9 8 7 6 5 4 3 2

ISBN 978-0-00-849210-6

British Library Cataloguing-in-Publication Data
A catalogue record for this publication is available from the British Library.

Authors: Lucy Wood, Jeremy Pollard, Sarah Jinks and Dorothy Warren
Publisher: Katie Sergeant
Product manager: Joanna Ramsay
Product developer: Holly Woolnough
Development editor: Rebecca Ramsden
Copyeditor: Jan Schubert
Proofreader: Helius
Cover designer: Kneath Associates
Cover illustrations: Mialapi/Shutterstock
Internal designer: Checkpoint
Production controller: Alhady Ali

For more information visit:
www.harpercollins.co.uk/green
Acknowledgements

The publishers gratefully acknowledge the permission granted to reproduce the copyright material in this book. Every effort has been made to trace copyright holders and to obtain their permission for the use of copyright material. The publishers will gladly receive any information enabling them to rectify any error or omission at the first opportunity.

Printed and bound in the UK using 100% Renewable Electricity at CPI Group (UK) Ltd

FSC
www.fsc.org
MIX
Paper from responsible sources
FSC™ C007454

This book is produced from independently certified FSC™ paper to ensure responsible forest management.

Table of Contents

INTRODUCTION AND SEQUENCING

A note from the author

I am extremely excited by the release of Secure Science for GCSE. My passion is to help as many young people as possible understand the big science ideas, partly so they achieve in exams so they can unlock the next phase of their learning, and partly so they understand the science they encounter in their lives. I'm sure you agree that the latter aim has never been so important! In 20 years in science classrooms, I have often felt the tension between the natural connections that exist between science topics and the order in which we often feel compelled to teach those ideas. Secure Science for GCSE has been mapped to cover all the ideas students need for their exams in an order that allows for rapid progression through naturally connected concepts. Students are encouraged to structure their thinking using the carefully thought-through questions, and to use the images to aid understanding and memory. This is a course full of scientific connections and the tools that students need to understand science. I am incredibly proud to have joined the Collins team in the journey of its creation. I hope you enjoy using it as much as I enjoyed writing it.

How it works: THINK>CONNECT>SOLVE>PRACTISE

Secure Science for GCSE is an evidence-based programme designed to help students fix gaps, make connections, and build and secure a deep understanding of foundation-tier topics for GCSE Science 9-1. We've put together a programme in 30 sessions, which we hope will be of real value to students who need extra support, whether alongside timetabled teaching, as a part of an intervention programme or in advance of their exams. This introduction provides brief details of the overall structure, and we hope it will be useful to you in your planning and to get Secure Science implemented quickly in your school. We hope that you and your students find this resource supportive, confidence-building and effective.

THINK

Students start with the exam-style question for the session in the workbook. There is structured encouragement to retrieve knowledge in the writing frame. The key idea and apparatus cards at the back of the Teacher Resource can be printed out so that the student can deal out the relevant cards and use those to help them write down some notes. Students will not necessarily fill the space given in the workbook. Explain to them that the point of the exercise is to get their minds working and thinking about the topic, and this is all helping to activate and strengthen their memory.

Corresponding modules for each Secure Science session are hosted on the Adapt© platform, and these can be set as assignments for students by their teacher. Students will see a range of questions that require them to recall their knowledge on that topic. Critically, they will also be asked to rate their confidence level for each question: the questions will then adapt according to each student's competence and confidence levels, so that they create their own pathway towards mastery. There is immediate, specific feedback on each question, and relevant explainer content is shown to students to help them. The data on student competence can be used to drill down to particular misconceptions or topics that the student finds challenging, and guide your onward teaching and support.

CONNECT

Having completed the questions, the student can watch a video on Adapt© about the powerful idea underpinning the session topic, or look at the knowledge organiser in the workbook, which summarises the key information and how it links together. This will help the student to make connections between topics and understand underlying principles.

SOLVE

Now the student can return to the workbook to solve the session question. The supportive writing frame will help the student to build confidence and skills in answering questions, identifying and understanding key vocabulary, working out which ideas are relevant and learning how to structure an answer. They can still have the key idea cards dealt out beside them for support.

Look out for misconceptions if you are discussing the question with the student. A list of common misconceptions and how to address them is given in the Teacher Resource.

We have provided sample answers with commentary for two questions from each science for additional support and scaffolding for sessions 1, 9, 13, 19, 25 and 29. Students can work through these if they would benefit from extra help before tackling exam questions on their own.

PRACTISE

Secure Science offers plenty of opportunities for interleaving and repeated practice. The sessions on Adapt© can be repeated where students want to check and strengthen their knowledge. As the sessions adapt to students' competence and confidence levels, they remain fresh and are always supportive. Students can re-watch the explainer videos on Adapt© for revision or consolidation.

There are additional practice questions for each session and for all command words that require a longer response. Students can use these independently and we have provided sequencing suggestions if you want to build them into your teaching. Finally there are generic writing frames at the end of the workbook that can be used with any exam-style question using a relevant command word. This gives students strong support for all their practice and revision sessions.

Curriculum structure rationale

The curriculum structure is based on our experience of how expert teachers enable pupils to cover large areas of content and do well at GCSE.

We have mapped out the key topics that pupils need for GCSE science, and tracked back to the core concepts from Key Stage 3 that make the key topics make sense. Students often have gaps in their prior learning, and this forms a barrier to them grasping the concepts required for GCSE. This condensed curriculum includes the relevant Key Stage 3 knowledge alongside the GCSE topics so that students can secure their understanding of the basic concepts and so that they have a firm foundation on which to build the understanding required for success in GCSE science. Students who do well in GCSE tend to have a well-developed understanding of the basic concepts at Key Stage 3, and there are questions in the exam papers that need only a Key Stage 3 understanding. Each session has a problem that needs pupils to understand the whole concept. The structured writing frame enables pupils to think through the basic concept before making links to the GCSE content and constructing an answer where the whole concept is applied to solve the problem.

Flexibility

The resources can be used flexibly to meet the needs of your students. Time spent on different elements of the session will vary. For example, time spent answering questions on Adapt© will vary depending on how confident a student is about a topic and how secure their knowledge is. If the topic knowledge is secure, then the student can spend more time applying what they have learnt by solving the exam-style question in the workbook.

Secure Science can be taught in class, used with tutors one-to-one or students can use the resource independently. If teaching students in class, then the misconceptions sections and the sample answers with commentary in the Teacher Resource can support any re-teaching you would like to do. If a student is working independently, then they can re-watch video explainers if they want to consolidate knowledge and use the key idea cards to give them prompts about material to include in an answer.

Sequencing suggestions

Secure Science can be used in several ways as students progress through years 10 and 11.

1. Teaching throughout the GCSE course

We know that timetabling science is complicated, and students are often taught by more than one teacher. Secure Science sessions can be used after each of the GCSE units has been taught so that the practice questions help students recall the concepts they need for later units, before teaching those units. That way, the earlier Secure Science sessions can be used if students have gaps in their prior learning.

The Secure Science practical sessions are organised by discipline. Practicals relevant to each unit are shown. The sequence below suggests how Secure Science sessions and practice questions link to the AQA GCSE Combined Science (Trilogy) topics.

AQA topic and examination paper	Secure Science unit to consolidate learning after teaching	Secure Science practice questions to assess prior knowledge before teaching	Practical science session
1. Cell biology Biology Paper 1	1. Biological structures and their functions 2. Moving molecules in living things		
2. Organisation Biology Paper 1	3. Life systems 4. Staying alive	1. Biological structures and their functions	
3. Infection and response Biology Paper 1	7. Disease and immunity	1. Biological structures and their functions	
4. Bioenergetics Biology Paper 1	8. Life cycles	1. Biological structures and their functions	11. Biology practicals
5. Homeostasis and response Biology Paper 2	5. Nervous control 6. Hormonal control	2. Moving molecules in living things	
6. Inheritance, variation and evolution Biology Paper 2	9. Reproduction and inheritance 10. Genetics and genetic engineering	1. Biological structures and their functions	
7. Ecology Biology Paper 2	8. Life cycles	2. Moving molecules in living things	
8. Atomic structure and the periodic table Chemistry Paper 1	13. Atomic structure 17 Elements, mixture and compounds		
9. Bonding, structure and the properties of matter Chemistry Paper 1	12. The particle model 16. Properties of materials	13. Atomic structure	
10. Quantitative chemistry Chemistry Paper 1	14. Calculating chemical change	13. Atomic structure	22. Chemistry practicals
11. Chemical changes Chemistry Paper 1	14. Calculating chemical change 15. Bonding 19. Electrolysis 20. Acids and alkalis	12. The particle model	

AQA topic and examination paper	Secure Science unit to consolidate learning after teaching	Secure Science practice questions to assess prior knowledge before teaching	Practical science session
12. Energy changes Chemistry Paper 1	21. Rates of reaction	12. The particle model	
13. The rate and extent of chemical change Chemistry Paper 2	21. Rates of reaction	14. Calculating chemical change	
14. Organic chemistry Chemistry Paper 2	18. Earth chemistry	12. The particle model	
15. Chemical analysis Chemistry Paper 2	14. Calculating chemical change	13. Atomic structure	
16. Chemistry of the atmosphere Chemistry Paper 2	18. Earth chemistry	14. Calculating chemical change	
17. Using resources Chemistry Paper 2	18. Earth chemistry	3. Life systems	
18. Energy Physics Paper 1	26. Energy	13. Atomic structure	30. Physics practicals
19. Electricity Physics Paper 1	27. Electricity	13. Atomic structure	
20. Particle model of matter Physics Paper 1	12. The particle model 29. Atoms and radioactivity	12. The particle model	
21. Atomic structure Physics Paper 1	29. Atoms and radioactivity	13. Atomic structure	
22. Forces Physics Paper 2	23. Forces 24. Motion	12. The particle model	
23. Waves Physics Paper 2	25. Waves	12. The particle model	
24. Magnetism and electromagnetism Physics Paper 2	28. Electromagnetism	16. Properties of materials	

2. 12-week model growing students' confidence after mocks and before the exam

Secure Science sessions can support students to build their understanding of the big concepts of science at the end of year 11, when teaching of the specification finishes in the spring term. A suggested route through for teaching in lessons is given below.

Block 1: Underpinning concepts
1. Biological structures and their functions
2. Moving molecules in living things
12. The particle model
17. Elements, compounds and mixtures
13. Atomic structure
Block 2: Using the concepts
3. Life systems
4. Staying alive
7. Disease and immunity
8. Life cycles
14. Calculating chemical change
15. Bonding
20. Acids and alkalis
16. Properties of materials
23. Forces
25. Waves
26. Energy
Block 3: Applying the concepts
29. Atoms and radioactivity
28. Electromagnetism
27. Electricity
24. Motion
10. Genetics and genetic engineering
9. Reproduction and inheritance
7. Disease and immunity
6. Hormonal control
5. Nervous control
14. Calculating chemical change
18. Earth chemistry
19. Electrolysis
21. Rates of reaction
30. Physics practicals
22. Chemistry practicals
11. Biology practicals

3. 12-week model with interleaved practice questions

The following table shows a suggested route through for learning in taught lessons and private study, either at home or in revision lessons after school. This model is taught over 12 weeks, assuming students will be studying from the end of their mocks in February to the end of May. The taught sessions are in order of conceptual development — in other words, the concepts taught at the beginning need to be covered first so that students can access the content in the subsequent units.

The interleaved content grows as students progress through the units, as they need to be practicing everything they have learned so far. Interleaving relies on the revisiting happening and the quality of questions, rather than the number of questions, so choosing one or two key questions is more effective than trying to cover all the previous content.

During teaching of Block 1 sessions, the additional practice questions could be used a short while after each session to revisit the ideas and grow students' confident in accurate recall of the ideas.

During teaching of Block 2 sessions, the Block 1 session additional practice questions can be used for students to practice recalling the previous concepts. During teaching of Block 3 sessions, the additional practice questions from Blocks 1 and 2 should be used. Practicing the recall of prior knowledge and equations can be done either in taught lessons, during revision sessions or at home.

Week	Block	Taught sessions	Interleaved additional practice questions
1	1	1. Biological structures and their functions 12. The particle model	1. Biological structures and their functions 12. The particle model
2		2. Moving molecules in living things 17. Elements, compounds and mixtures 13. Atomic structure	2. Moving molecules in living things 17. Elements, compounds and mixtures 13. Atomic structure
3	2	3. Life systems 15. Bonding 23. Forces	1. Biological structures and their functions 17. Elements, compounds and mixtures
4		4. Staying alive 20. Acids and alkalis 25. Waves	2. Moving molecules in living things 17. Elements, compounds and mixtures
5		7. Disease and immunity 16. Properties of materials 26. Energy	4. Staying alive 15. Bonding 13. Atomic structure 23. Forces
6.		8. Life cycles 14. Calculating chemical change 11. Biology practicals	3. Life systems 13. Atomic structure 20. Acids and alkalis 25. Waves
7.	3	24. Motion 6. Hormonal control 5. Nervous control	4. Staying alive 12. The particle model 23. Forces 26. Energy
8		29. Atoms and radioactivity 18. Earth chemistry 30. Physics practicals	6. Hormonal control 13. Atomic structure 15. Bonding

Week	Block	Taught sessions	Interleaved additional practice questions
			16. Properties of materials 20. Acids and alkalis 25. Waves
9		28. Electromagnetism 7. Disease and immunity 14. Calculating chemical change	4. Staying alive 11. Biology practicals 13. Atomic structure 18. Earth chemistry 24. Motion 26. Energy
10		9. Reproduction and inheritance 19. Electrolysis	5. Nervous control 8. Life cycles 12. The particle model 14. Calculating chemical change 28. Electromagnetism 29. Atoms and radioactivity 30. Physics practicals
11		27. Electricity 10. Genetics and genetic engineering	6. Hormonal control 7. Disease and immunity 8. Life cycles 16. Properties of materials 17. Elements, compounds and mixtures 20. Acids and alkalis
12		21. Rates of reaction 22. Chemistry practicals	5. Nervous control 9. Reproduction and inheritance 11. Biology practicals 14. Calculating chemical change 18. Earth chemistry 19. Electrolysis 24. Motion 26. Energy 27. Electricity 29. Atoms and radioactivity
Revisit later			9. Reproduction and inheritance 10. Genetics and genetic engineering 19. Electrolysis 21. Rates of reaction 22. Chemistry practicals 23. Forces 24. Motion 25. Waves 27. Electricity 30. Physics practicals 11. Biology practicals

4. An accelerated approach

If time is short, it is vital to cover the basic concepts that are both easy to access and unlock a lot of science knowledge.

Some of the Secure Science sessions form the backbone of a basic understanding of science, and can provide the maximum amount of valuable science learning in just ten sessions:

Session number	Secure Science session title
1	Biological structures and their functions
2	Moving molecules in living things
3	Life systems
12	The particle model
13	Atomic structure
17	Elements, compounds and mixtures
23	Forces
25	Waves
27	Electricity
22	Chemistry practicals

5. Independent student revision programme immediately before exams

Students on study leave or working independently can use Secure Science as a revision resource immediately before exams in May or June. The table below indicates which Secure Science sessions are relevant to each examination paper. Organised students could use these sessions to cover the key concepts independently before each examination.

Examination paper	Relevant Secure Science session
Biology Paper 1	1. Biological structures and their functions 2. Moving molecules in living things 3. Life systems 4. Staying alive 7. Disease and immunity 8. Life cycles
Biology Paper 2	5. Nervous control 6. Hormonal control 8. Life cycles 9. Reproduction and inheritance 10. Genetics and genetic engineering

Examination paper	Relevant Secure Science session
Chemistry Paper 1	12. The particle model 13. Atomic structure 14. Calculating chemical change 15. Bonding 16. Properties of materials 17. Elements, compounds and mixtures 19. Electrolysis 20. Acids and alkalis 21. Rates of reaction
Chemistry Paper 2	14. Calculating chemical change 18. Earth chemistry 21. Rates of reaction
Physics Paper 1	12. The particle model 26. Energy 27. Electricity 29. Atoms and radioactivity
Physics Paper 2	23. Forces 24. Motion 25. Waves 28. Electromagnetism

MAPPING TO AQA GCSE 9-1 AND KS3 SCIENCE

Secure Science for GCSE can be used with any exam board. We have mapped to the AQA GCSE Combined Science specification as this is most frequently used.

Session	Key ideas	AQA GCSE Combined Science specification	KS3 Programme of Study
1. Biological structures and their functions	• Specialised human cells • Specialised plant cells • Plant and animal cells • Cell structures	4.1.1.1 Eukaryotes and prokaryotes [Bacterial cells (prokaryotic cells) not covered until Session 7; use of standard form not covered] 4.1.1.2 Animal and plant cells [Use of estimations not covered] 4.1.1.3 Cell specialisation	• cells as the fundamental unit of living organisms, [including how to observe, interpret and record cell structure using a light microscope] • the functions of the cell wall, cell membrane, cytoplasm, nucleus, vacuole, mitochondria and chloroplasts • the similarities and differences between plant and animal cells • the adaptations of leaves for photosynthesis
2. Moving molecules in living things	• Diffusion • Osmosis • Moving water through a plant • Diffusion in plants • Osmosis in animal cells • The structure of plants	4.1.3.1 Diffusion [Diffusion of urea not covered; factors affecting rate of diffusion not covered; single-celled organism having large surface area to volume ratio not covered; calculating; comparing and explaining surface area to volume ratio not covered; adaptations of small intestine and lungs for exchange not covered until Session 3; effectiveness of exchange surface not covered until Session 3] 4.1.3.2 Osmosis	• the role of diffusion in the movement of materials in and between cells • the role of leaf stomata in gas exchange in plants • diffusion in liquids and gases driven by differences in concentration
3. Life systems	• Lungs and gas exchange • The digestive	4.1.3.1 Diffusion [Adaptations of gills in fish not covered] 4.2.2.1 The human digestive	• the role of diffusion in the movement of materials in and between cells • the hierarchical organisation of multicellular organisms: from cells to tissues to organs to

Session	Key ideas	AQA GCSE Combined Science specification	KS3 Programme of Study
	system • Absorption in the small intestine • Enzyme action • Enzymes and digestion	system [Enzyme activity related to temperature and pH changes not covered; rate calculations for chemical reactions not covered; word equations not covered; production and function of bile not covered]	systems to organisms • content of a healthy human diet: carbohydrates, lipids (fats and oils), proteins, vitamins, minerals, dietary fibre and water, and why each is needed • the tissues and organs of the human digestive system, including adaptations to function and how the digestive system digests food (enzymes simply as biological catalysts) • the structure and functions of the gas exchange system in humans, including adaptations to function • [calculations of energy requirements in a healthy daily diet not covered] • [the consequences of imbalances in the diet, including obesity, starvation anddeficiency diseases not covered] • [the importance of bacteria in the human digestive system not covered]
4. Staying alive	• The respiratory system • The circulatory system • Structure of the heart • Blood and blood vessels	4.2.1 Principles of organisation 4.2.2.2 The heart and blood vessels [Natural and artificial pacemaker not covered; rate calculations not covered] 4.2.2.3 Blood [Appearance and adaptations of blood cells not covered] 4.2.2.4 Coronary heart disease: a non-communicable disease [Treatment not covered; faulty valves not covered] 4.2.2.6 The effect of lifestyle on	• the hierarchical organisation of multicellular organisms: from cells to tissues to organs to systems to organisms • the structure and functions of the gas exchange system in humans, including adaptations to function • [the mechanism of breathing to move air in and out of the lungs, using a pressure modelto explain the movement of gases, including simple measurements of lung volume not covered] • [the impact of exercise, asthma and smoking on the human gas exchange system not covered] • [the effects of recreational drugs (including substance misuse) on behaviour, health andlife

Session	Key ideas	AQA GCSE Combined Science specification	KS3 Programme of Study
		some non-communicable diseases [Human and financial cost not covered; effects of diet, alcohol, smoking and obesity not covered; carcinogens not covered; use of scatter diagram to identify a correlation not covered]	processes not covered]
5. Nervous control	• The nervous system • Structure of nerve cells • Coordinating responses • The reflex arc	4.5.2 The human nervous system	• the hierarchical organisation of multicellular organisms: from cells to tissues to organs to systems to organisms • [the structure and functions of the human skeleton, to include support, protection,movement and making blood cells not covered] • [biomechanics – the interaction between skeleton and muscles, including themeasurement of force exerted by different muscles not covered] • [the function of muscles and examples of antagonistic muscles not covered]
6. Hormonal control	• The endocrine system • Homeostasis • Controlling blood sugar • Puberty	4.5.1 Homeostasis 4.5.3.1 Human endocrine system 4.5.3.2 Control of blood glucose concentration 4.5.3.3 Hormones in human reproduction [Testosterone as main male reproductive hormone not covered] 4.5.3.4 Contraception [Intrauterine devices, spermicidal agents, abstention and sterilisation not covered]	• the hierarchical organisation of multicellular organisms: from cells to tissues to organs to systems to organisms • reproduction in humans (as an example of a mammal), including the structure and function of the male and female reproductive systems, menstrual cycle (without details of hormones), gametes, fertilisation, gestation and birth, to include the effect of maternal lifestyle on the foetus through the placenta

Session	Key ideas	AQA GCSE Combined Science specification	KS3 Programme of Study
7. Disease and immunity	• The immune system • Preventing disease • Pathogens • Pathogens and their diseases	4.1.1.1 Eukaryotes and prokaryotes 4.3.1.1 Communicable (infectious) diseases 4.3.1.2 Viral diseases 4.3.1.3 Bacterial diseases [Prevention of gonorrhoea by using condom not covered] 4.3.1.4 Fungal diseases [Treatment using fungicides not covered] 4.3.1.5 Protist diseases [Use of mosquito nets not covered] 4.3.1.6 Human defence systems 4.3.1.7 Vaccination [Spread of pathogens being reduced by immunising large proportion of population not covered] 4.3.1.8 Antibiotics and painkillers [Painkillers not covered; difficulty of developing drugs that kill viruses not covered]	• the structural adaptations of some unicellular organisms
8. Life cycles	• Photosynthesis • Respiration • Ecosystems • Nutrient cycling • Human interactions	4.4.1.1 Photosynthetic reaction [Photosynthesis being an endothermic reaction not covered] 4.4.1.3 Uses of glucose from photosynthesis [Use of nitrate ions from soil not covered] 4.4.2.1 Aerobic and anaerobic respiration	• plants making carbohydrates in their leaves by photosynthesis and gaining mineral nutrients and water from the soil via their roots • the reactants in, and products of, photosynthesis, and a word summary for photosynthesis • the dependence of almost all life on Earth on the ability of photosynthetic organisms, such as plants and algae, to use sunlight in photosynthesis to build organic molecules that are an essential

Session	Key ideas	AQA GCSE Combined Science specification	KS3 Programme of Study
		[Respiration being an exothermic reaction not covered; what organisms need energy for not covered] 4.7.1.1 Communities [Organisms being adapted to the conditions in which they live not covered until Session 9; stable community not covered] 4.7.2.1 Levels of organisation [Determination by ecologists of distribution and abundance of species not covered; terms mean, mode, median not covered; plotting of graphs not covered] 4.7.2.2 How materials are cycled 4.7.3.1 Biodiversity [Advantages of a great biodiversity not covered] 4.7.3.2 Waste management [Where pollution can occur, and its impact on plants and animals, not covered] 4.7.3.4 Deforestation [Reasons for large-scale deforestation not covered] 4.7.3.5 Global warming [Biological consequences of global warming not covered] 4.7.3.6 Maintaining biodiversity [Reintroduction of field margins and hedgerows, reduction of	energy store and to maintain levels of oxygen and carbon dioxide in the atmosphere • the adaptations of leaves for photosynthesis • aerobic and anaerobic respiration in living organisms, including the breakdown of organic molecules to enable all the other chemical processes necessary for life • a word summary for aerobic respiration • the process of anaerobic respiration in humans and micro-organisms, including fermentation, and a word summary for anaerobic respiration • the differences between aerobic and anaerobic respiration in terms of the reactants, the products formed and the implications for the organism • the interdependence of organisms in an ecosystem, including food webs and insect pollinated crops • the importance of plant reproduction through insect pollination in human food security • how organisms affect, and are affected by, their environment, including the accumulation of toxic materials • the importance of maintaining biodiversity [and the use of gene banks to preserve hereditary material not covered]

Session	Key ideas	AQA GCSE Combined Science specification	KS3 Programme of Study
		deforestation and carbon dioxide emissions, recycling resources not covered]	
9. Reproduction and inheritance	• Meiosis • Sexual reproduction • Evolution by natural selection • Variation	4.1.2.1 Chromosomes [Nucleus of cell containing chromosomes made up of DNA and carrying a large number of genes not covered until Session 10] 4.6.1.1 Sexual and asexual reproduction [Mitosis not covered until Session 10] 4.6.1.2 Meiosis [Number of cells increasing and differentiating as embryo develops not covered until Session 10] 4.6.1.6 Sex determination [Human body cells containing 23 pairs of chromosomes not covered until Session 10] 4.6.2.1 Variation 4.6.2.2 Evolution [All species having evolved from simple life forms more than 3 billion years ago not covered] 4.6.3.1 Evidence for evolution [Evolution of antibiotic resistance in bacteria not covered] 4.7.1.4 Adaptations [Extremophiles not covered]	• reproduction in humans (as an example of a mammal), including the structure and function of the male and female reproductive systems, menstrual cycle (without details of hormones), gametes, fertilisation, gestation and birth, to include the effect of maternal lifestyle on the foetus through the placenta • reproduction in plants, including flower structure, wind and insect pollination, fertilisation, seed and fruit formation and dispersal, including quantitative investigation of some dispersal mechanisms • the variation between species and between individuals of the same species means some organisms compete more successfully, which can drive natural selection • [differences between species not covered] • [the variation between individuals within a species being continuous or discontinuous,to include measurement and graphical representation of variation not covered] • [changes in the environment may leave individuals within a species, and some entire species, less well adapted to compete successfully and reproduce, which in turnmay lead to extinction not covered]
10. Genetics and genetic	• Mitosis and stem	4.1.1.4 Cell differentiation	• heredity as the process by which genetic

Session	Key ideas	AQA GCSE Combined Science specification	KS3 Programme of Study
engineering	cells • DNA • Genes • Genetics • Genetic engineering	[Plant cell differentiation not covered] 4.1.2.1 Chromosomes 4.1.2.2 Mitosis and the cell cycle [Cell growing and increasing number of subcellular structures before division not covered; DNA replicating not covered; chromosomes being pulled to each end of cell and nucleus dividing not covered; cytoplasm and cell membranes dividing not covered] 4.1.2.3 Stem cells [Meristem cells in plants not covered; potential risks of stem cell treatment not covered] 4.6.1.3 DNA and the genome [The importance of studying the human genome not covered] 4.6.1.4 Genetic inheritance [Some characteristics being controlled by a single gene but most characteristics being the result of multiple genes not covered] 4.6.1.5 Inherited disorders [Economic, social and ethical issues concerning embryo screening not covered] 4.6.2.3 Selective breeding [Characteristics chosen for selective breeding not covered; selective breeding leading to	information is transmitted from one generation to the next • a simple model of chromosomes, genes and DNA in heredity [including the part playedby Watson, Crick, Wilkins and Franklin in the development of the DNA model not covered]

Session	Key ideas	AQA GCSE Combined Science specification	KS3 Programme of Study
		'inbreeding' not covered] 4.6.2.4 Genetic engineering [Genetic engineering of bacterial cells to produce human insulin not covered]	
11. Biology practicals	• Making a slide • Calculating magnification of an image • Unit conversion • Calculating total magnification • Using a microscope • Investigating osmosis • Investigating molecules in foods • Investigating rate of reaction • Investigating photosynthesis • Investigating reaction time • Measuring population size • Ethics in biology experiments	Required practical activity 1: Use a light microscope Required practical activity 2: Investigate the effect of a range of concentration of salt [or sugar] on mass of plant tissue Required practical activity 3: Test for a range of carbohydrates, lipids and proteins Required practical activity 4: Investigate the effect of pH on the rate of reaction of amylase enzyme Required practical activity 5: Investigate the effect of light intensity on the rate of photosynthesis Required practical activity 6: [Plan and] carry out an investigation into the effect of a factor on human reaction time Required practical activity 7: Measure the population size of a common species in a habitat 4.1.1.5 Microscopy [Development of microscopy techniques and electron microscopy not covered; resolution not covered; use of standard form	• pay attention to objectivity and concern for accuracy, precision, repeatability and reproducibility • understand that scientific methods and theories develop as earlier explanations are modified to take account of new evidence and ideas, together with the importance ofpublishing results and peer review • evaluate risks • ask questions and develop a line of enquiry based on observations of the real world,alongside prior knowledge and experience • make predictions using scientific knowledge and understanding • select, plan and carry out the most appropriate types of scientific enquiries to test predictions, including identifying independent, dependent and control variables, whereappropriate • use appropriate techniques, apparatus, and materials during fieldwork and laboratorywork, paying attention to health and safety • make and record observations and measurements using a range of methods for different investigations; and evaluate the reliability of methods and suggest possibleimprovements • apply sampling techniques. • apply mathematical concepts and calculate results

Session	Key ideas	AQA GCSE Combined Science specification	KS3 Programme of Study
		not covered] 4.4.1.2 Rate of photosynthesis [Effects of temperature, carbon dioxide concentration, and amount of chlorophyll not covered]	• present observations and data using appropriate methods, including tables and graphs • interpret observations and data, including identifying patterns and using observations, measurements and data to draw conclusions • present reasoned explanations, including explaining data in relation to predictions and hypotheses • evaluate data, showing awareness of potential sources of random and systematic error • identify further questions arising from their results
Biology sections not covered	4.1.3.3, 4.2.2.5, 4.2.2.7, 4.2.3.1, 4.2.3.2, 4.3.1.9, 4.4.2.2, 4.4.2.3, 4.6.3.2, 4.6.3.3, 4.6.3,4, 4.6.4, 4.7.1.2, 4.7.1.3		
12. The particle model	• Particle model • Changing state • Density • Internal energy	5.2.2.1 The three states of matter [Predicting states of substances at different temperatures not covered; recognising that atoms do not have the bulk properties of the material not covered] 6.3.1.1 Density of materials 6.3.1.2 Changes of state [Conservation of mass not covered until Session 14; changes of state as physical changes not covered] 6.3.2.1 Internal energy 6.3.2.2 Temperature changes in a system and specific heat capacity	• the properties of the different states of matter (solid, liquid and gas) in terms of the particle model [including gas pressure] • changes of state in terms of the particle model • energy changes on changes of state (qualitative) • [diffusion in terms of the particle model not covered] • [conservation of material and of mass not covered], and reversibility, in melting, freezing,evaporation, sublimation, condensation, dissolving • similarities and differences, including density differences, between solids, liquids and gases • [Brownian motion in gases not covered] • the difference between chemical and physical changes • the differences in arrangements, in motion and in closeness of particles explaining changes of state, shape and density, the anomaly of ice-water

Session	Key ideas	AQA GCSE Combined Science specification	KS3 Programme of Study
			transition • atoms and molecules as particles • changes with temperature in motion and spacing of particles • internal energy stored in materials
13. Atomic structure	• Atomic structure • Formula • Periodic table • Periodicity	5.1.1.1 Atoms, elements and compounds [Compounds being formed from elements by chemical reactions not covered until Session 14; compounds only being separated into elements by chemical reactions not covered; word equations and formula equations for chemical reactions not covered until Session 14] 5.1.1.4 Relative electrical charges of subatomic particles 5.1.1.5 Size and mass of atoms [Radius of atom and of nucleus not covered; isotopes not covered] 5.1.1.7 Electronic structure 5.1.2.1 The periodic table [Prediction of possible reactions not covered] 5.1.2.3 Metals and non-metals [Positions on periodic table only; remainder covered in Session 15] 5.1.2.4 Group 0 [Position on periodic table and electronic structure only]	• a simple (Dalton) atomic model • differences between atoms, elements and compounds • chemical symbols and formulae for elements and compounds • chemical reactions as the rearrangement of atoms • the varying physical and chemical properties of different elements • the principles underpinning the Mendeleev periodic table • the periodic table: periods and groups; metals and non-metals • how patterns in reactions can be predicted with reference to the periodic table • [combustion, thermal decomposition, oxidation and displacement reactions not covered]

Session	Key ideas	AQA GCSE Combined Science specification	KS3 Programme of Study
		5.1.2.5 Group 1 [Name alkali metals not covered; descriptions of reactions not covered] 5.1.2.6 Group 7 [Reactivity going down the group only]	
14. Calculating chemical change	• Chemical and physical change • Tests for common gases • Word and symbol equations • Conservation of mass • Calculating chemical quantities	5.1.1.1 Atoms, elements and compounds 5.3.1.1 Conservation of mass and balanced chemical equations 5.3.1.2 Relative formula mass 5.3.1.3 Mass changes when a reactant or product is a gas 5.8.2.1 Test for hydrogen 5.8.2.2 Test for oxygen 5.8.2.3 Test for carbon dioxide 5.8.2.4 Test for chlorine	• chemical symbols and formulae for elements and compounds • conservation of mass, changes of state and chemical reactions • chemical reactions as the rearrangement of atoms • representing chemical reactions using formulae and using equations • conservation of material and of mass, [and reversibility, in melting, freezing,evaporation, sublimation, condensation, dissolving not covered]
15. Bonding	• Ionic bonding • Covalent bonding • Type of bonding • Metallic bonding	5.1.2.3 Metals and non-metals 5.2.1.1 Chemical bonds 5.2.1.2 Ionic bonding 5.2.1.3 Ionic compounds 5.2.1.4 Covalent bonding [Giant covalent structures not covered until Session 16; diagrammatic/formulaic representations of covalently bonded molecules limited to dot and cross diagrams] 5.2.1.5 Metallic bonding	• chemical symbols and formulae for elements and compounds • the properties of metals and non-metals • properties of [ceramics,] polymers [and composites] (qualitative)

Session	Key ideas	AQA GCSE Combined Science specification	KS3 Programme of Study
		5.2.2.5 Polymers	
16. Properties of materials	• Small molecules • Giant covalent structures • Ionic compounds • Metals and alloys	5.2.1.4 Covalent bonding 5.2.2.1 The three states of matter 5.2.2.3 Properties of ionic compounds 5.2.2.4 Properties of small molecules [Intermolecular forces increasing with the size of the molecules not covered] 5.2.2.6 Giant covalent structures 5.2.2.7 Properties of metals and alloys 5.2.2.8 Metals as conductors [Metals as good conductors of thermal energy not covered] 5.2.3.1 Diamond 5.2.3.2 Graphite	• the concept of a pure substance • mixtures [including dissolving] • the properties of metals and non-metals
17. Elements, compounds and mixtures	• Elements, compounds and mixtures • Separation techniques • Chromatography • Pure and impure substances	5.1.1.2 Mixtures 5.8.1.1 Pure substances 5.8.1.3 Chromatography [Providing answers to an appropriate number of significant figures not covered] Required practical 12: Investigate how paper chromatography can be used to separate and tell the difference between coloured substances, and calculate R_f values	• differences between atoms, elements and compounds • the concept of a pure substance • mixtures, including dissolving • simple techniques for separating mixtures: filtration, evaporation, distillation and chromatography • the identification of pure substances
18. Earth chemistry	• Hydrocarbons	5.2.2.5 Polymers	• the carbon cycle

Session	Key ideas	AQA GCSE Combined Science specification	KS3 Programme of Study
	• Fractional distillation • Useful materials • Earth's atmosphere • Polymers	5.7.1.1 Crude oil, hydrocarbons and alkanes 5.7.1.2 Fractional distillation and petrochemicals 5.7.1.3 Properties of hydrocarbons [Boiling points, viscosity and flammability of hydrocarbons depending on the size of their molecules not covered; oxidation of carbon and hydrogen in combustion of fuels not covered] 5.7.1.4 Cracking and alkanes [Methods of and conditions for cracking not covered; reactivity and test for alkenes not covered] 5.9.1.1 The proportions of different gases in the atmosphere [Proportions being much the same for 200 million years not covered] 5.9.1.2 The Earth's early atmosphere [Timescale not covered; comparison with Mars and Venus not covered; production of nitrogen, methane and ammonia not covered] 5.9.1.3 How oxygen increased [Formula equation for photosynthesis not covered; timescale not covered] 5.9.1.4 How carbon dioxide decreased [Formation of sedimentary rocks	• the composition of the atmosphere • [the composition of the Earth not covered] • [the structure of the Earth not covered] • [the rock cycle and the formation of igneous, sedimentary and metamorphic rocks not covered] • [Earth as a source of limited resources and the efficacy of recycling not covered] • [the production of carbon dioxide by human activity and the impact on climate not covered]

Session	Key ideas	AQA GCSE Combined Science specification	KS3 Programme of Study
		and fossil fuels not covered]	
19. Electrolysis	• Extracting metals • Electrolysis • Conservation of mass • Electrolysis of aqueous solutions	5.3.1.1 Conservation of mass and balanced chemical equations [Balanced equations not covered] 5.4.1.2 The reactivity series [Reactivity being related to tendency to form positive ions not covered; displacement not covered; specific reactions not covered] 5.4.1.3 Extraction of metals and reduction [Oxidation of substances not covered] 5.4.3.1 The process of electrolysis 5.4.3.2 Electrolysis of molten ionic compounds 5.4.3.3 Using electrolysis to extract metals 5.4.3.4 Electrolysis of aqueous solutions	• conservation of mass, changes of state and chemical reactions • chemical reactions as the rearrangement of atoms • the order of metals and carbon in the reactivity series • the use of carbon in obtaining metals from metal oxides
20. Acids and alkalis	• Acids and alkalis • General equations for making salts • Neutralisation • Conservation of mass • Salts	5.4.2.1 Reactions of acids with metals 5.4.2.2 Neutralisation of acids and salt production 5.4.2.3 Soluble salts 5.4.2.4 The pH scale and neutralisation [State symbols in equations not covered] 5.3.1.1 Conservation of mass and balanced chemical equations	• chemical symbols and formulae for elements and compounds • conservation of mass changes of state and chemical reactions • chemical reactions as the rearrangement of atoms • representing chemical reactions using formulae and using equations • defining acids and alkalis in terms of neutralisation reactions • the pH scale for measuring acidity/alkalinity; and indicators

Session	Key ideas	AQA GCSE Combined Science specification	KS3 Programme of Study
		5.3.1.3 Mass changes when a reactant or product is a gas	• reactions of acids with metals to produce a salt plus hydrogen • reactions of acids with alkalis to produce a salt plus water • the chemical properties of metal and non-metal oxides with respect to acidity
21. Rates of reaction	• Collision theory • Rates of reaction • Taking a closer look at surface area • Reaction profiles • Reversible reactions	5.6.1.2 Factors which affect the rates of chemical reactions 5.6.1.3 Collision theory and activation energy 5.6.1.4 Catalysts 5.5.1.1 Energy transfer during exothermic and endothermic reactions [Temperature change of surroundings not covered; examples and applications of exothermic and endothermic reactions not covered] 5.5.1.2 Reaction profiles 5.6.2.1 Reversible reactions 5.6.2.3 Equilibrium	• the properties of the different states of matter (solid, liquid and gas) in terms of the particle model, including gas pressure • what catalysts do • exothermic and endothermic chemical reactions (qualitative)
22. Chemistry practicals	• Risk assessment • Preparing soluble salts • Electrolysis of aqueous solutions • Investigating temperature changes in reacting solutions	Required practical activity 8: Preparation of a pure, dry sample of a soluble salt Required practical activity 9: Investigate what happens when aqueous solutions are electrolysed Required practical activity 10: Investigate the variables that affect temperature changes in reacting solutions	• pay attention to objectivity and concern for accuracy, precision, repeatability and reproducibility • understand that scientific methods and theories develop as earlier explanations are modified to take account of new evidence and ideas, together with the importance ofpublishing results and peer review • evaluate risks • ask questions and develop a line of enquiry based

Session	Key ideas	AQA GCSE Combined Science specification	KS3 Programme of Study
	• Investigating rates of reaction • Analysis and purification of water	Required practical activity 11: Investigate how changes in concentration affect the rates of reactions [Only by a method involving measuring the volume of gas produced] Required practical activity 13: Analysis and purification of water samples from different sources 5.3.2.5 Concentration of solutions [Calculation of mass of solute not covered] 5.6.1.1 Calculating rates of reactions [Calculating mean rate not covered; units not covered]	on observations of the real world,alongside prior knowledge and experience • make predictions using scientific knowledge and understanding • select, plan and carry out the most appropriate types of scientific enquiries to test predictions, including identifying independent, dependent and control variables, whereappropriate • use appropriate techniques, apparatus, and materials during fieldwork and laboratorywork, paying attention to health and safety • make and record observations and measurements using a range of methods for different investigations; and evaluate the reliability of methods and suggest possibleimprovements • apply sampling techniques. • apply mathematical concepts and calculate results • present observations and data using appropriate methods, including tables and graphs • interpret observations and data, including identifying patterns and using observations, measurements and data to draw conclusions • present reasoned explanations, including explaining data in relation to predictions and hypotheses • evaluate data, showing awareness of potential sources of random and systematic error • identify further questions arising from their results
Chemistry sections not covered	5.1.1.3 [but this is covered in Physics Session 29], 5.1.1.6, 5.1.2.2, 5.2.2.2, 5.2.3.3, 5.3.1.4, 5.4.1.1, 5.6.2.2, 5.8.1.2, 5.9.2.1, 5.9.2.2, 5.9.2.3, 5.9.2.4, 5.9.3.1, 5.9.3.2, 5.10.1.1, 5.10.1.2, 5.10.1.3, 5.10.2.1, 5.10.2.2		

Session	Key ideas	AQA GCSE Combined Science specification	KS3 Programme of Study
23. Forces	• Forces and their effects • Contact and non-contact forces • Resultant forces and Newton's Second Law • Newton's First and Third Laws • Weight	6.5.1.2 Contact and non-contact forces [Force as a vector quantity not covered until Session 24] 6.5.1.3 Gravity [Weight acting at centre of mass not covered; proportionality of weight and mass not covered; measurement of weight with spring-balance not covered until Session 30] 6.5.1.4 Resultant forces 6.5.4.2.1 Newton's First Law 6.5.4.2.2 Newton's Second Law [Terms proportional and inversely proportional not used; estimation of speeds, accelerations and forces not covered] 6.5.4.2.3 Newton's Third Law	• forces as pushes or pulls, arising from the interaction between two objects • using force arrows in diagrams, adding forces in one dimension, balanced and unbalanced forces • [moment as the turning effect of a force not covered] • forces: [associated with deforming objects; stretching and squashing – springs;] with rubbing and friction between surfaces, with pushing things out of the way; resistance to motion of air and water • forces measured in newtons, [measurements of stretch or compression as force is changed] • [work done and energy changes on deformation not covered] • non-contact forces: gravity forces acting at a distance on Earth and in space, forces between magnets and forces due to static electricity • opposing forces and equilibrium: [weight held by stretched spring or supported on a compressed surface] • forces being needed to cause objects to stop or start moving, or to change their speed or direction of motion (qualitative only) • change depending on direction of force and its size • gravity force, weight = mass × gravitational field strength (*g*), on Earth g = 10 N/kg, [different on other planets and stars; gravity forces between Earth and Moon, and between Earth and Sun (qualitative only)] • [our Sun as a star, other stars in our galaxy, other galaxies not covered]

Session	Key ideas	AQA GCSE Combined Science specification	KS3 Programme of Study
			• [the seasons and the Earth's tilt, day length at different times of year, in differenthemispheres not covered] • [the light year as a unit of astronomical distance not covered] • [atmospheric pressure, decreases with increase of height as weight of air abovedecreases with height not covered] • [pressure in liquids, increasing with depth; upthrust effects, floating and sinking not covered] • [pressure measured by ratio of force over area – acting normal to any surface not covered]
24. Motion	• Speed, velocity and acceleration • Graphs of motion • Stopping distances • Terminal velocity • Scalar and vector quantities	6.5.1.1 Scalar and vector quantities 6.5.4.1.2 Speed [As a scalar quantity and equation only] 6.5.4.1.3 Velocity [Vector/scalar distinction between displacement and distance not covered] 6.5.4.1.4 The distance—time relationship 6.5.4.1.5 Acceleration [estimation of everyday accelerations not covered; equation $v^2 - u^2 = 2as$ not covered; value of acceleration of free fall not covered] 6.5.4.3.1 Stopping distance 6.5.4.3.2 Reaction time [That it is affected by tiredness,	• speed and the quantitative relationship between average speed, distance and time (speed = distance ÷ time) • the representation of a journey on a distance–time graph • [relative motion: trains and cars passing one another not covered]

Session	Key ideas	AQA GCSE Combined Science specification	KS3 Programme of Study
		drugs, alcohol, distractions only] 6.5.4.3.3 Factors affecting braking distance 1 [Estimation of braking/stopping distance not covered]	
25. Waves	• Waves • Properties of waves • The electromagnetic spectrum • Properties and uses of the electromagnetic spectrum	6.6.1.1 Transverse and longitudinal waves [That medium does not travel not covered] 6.6.1.2 Properties of waves [Period not covered; method to measure speed of sound waves not covered; method for speed of ripples in water covered in Session 30] 6.6.2.1 Types of electromagnetic waves [Our eyes detecting only visible light not covered] 6.6.2.2 Properties of electromagnetic waves 1 6.6.2.3 Properties of electromagnetic waves 2 [Generation/absorption of electromagnetic waves from changes in atoms/nuclei not covered; specific hazards not covered] 6.6.2.4 Uses and applications of electromagnetic waves	• waves on water as undulations which travel through water with transverse motion; [these waves can be reflected, and add or cancel – superposition] • [the similarities and differences between light waves and waves in matter not covered] • light waves travelling through a vacuum; speed of light • [the transmission of light through materials: absorption, diffuse scattering and specular reflection at a surface not covered] • use of ray model to explain [imaging in mirrors, the pinhole camera,] the refraction of light [and action of convex lens in focusing (qualitative); the human eye] • [light transferring energy from source to absorber leading to chemical and electricaleffects; photo-sensitive material in the retina and in cameras not covered] • [colours and the different frequencies of light, white light and prisms (qualitative only);differential colour effects in absorption and diffuse reflection not covered] • [frequencies of sound waves, measured in hertz (Hz); echoes, reflection and absorptionof sound not covered] • [sound needs a medium to travel, the speed of

Session	Key ideas	AQA GCSE Combined Science specification	KS3 Programme of Study
			sound in air, in water, in solids not covered] • [sound produced by vibrations of objects, in loud speakers, detected by their effects onmicrophone diaphragm and the ear drum; sound waves are longitudinal not covered] • [auditory range of humans and animals not covered] • [pressure waves transferring energy; use for cleaning and physiotherapy by ultra-sound;waves transferring information for conversion to electrical signals by microphone not covered]
26. Energy	• Energy stores • Energy transfers • Power and efficiency • Energy resources	6.1.1.1 Energy stores and systems [Descriptions of changes in energy stores in specific situations not covered; 'work done' as energy transfer not covered] 6.1.1.2 Changes in energy [Elastic potential energy in stretched spring not covered] 6.1.1.4 Power [Power as rate at which 'work is done' not covered; examples of power not covered] 6.1.2.1 Energy transfers in a system [Ways of reducing unwanted energy transfers not covered; thermal conductivity and rate of cooling of a building not covered] 6.1.2.2 Efficiency 6.1.3 National and global energy resources	• [comparing energy values of different foods (from labels) (kJ) not covered] • comparing power ratings of appliances in watts (W, kW) • comparing amounts of energy transferred (J, kJ, kW hour) • [domestic fuel bills, fuel use and costs not covered] • fuels and energy resources • [simple machines give bigger force but at the expense of smaller movement (and viceversa): product of force and displacement unchanged not covered] • [heating and thermal equilibrium: temperature difference between two objects leading toenergy transfer from the hotter to the cooler one, through contact (conduction) or radiation; such transfers tending to reduce the temperature difference: use of insulators not covered] • other processes that involve energy transfer: changing motion, dropping an object, [completing an electrical circuit, stretching a spring, metabolism

Session	Key ideas	AQA GCSE Combined Science specification	KS3 Programme of Study
		[Patterns and trends in use of energy resources not covered]	of food, burning fuels] • energy as a quantity that can be quantified and calculated; the total energy has the same value before and after a change • comparing the starting with the final conditions of a system and describing increases and decreases in the amounts of energy associated with movements, [temperatures,] changes in positions in a field, [in elastic distortions and in chemical compositions] • [using physical processes and mechanisms, rather than energy, to explain theintermediate steps that bring about such changes not covered]
27. Electricity	• Electrical circuits • Electrical power • Resistance • Resistance of a wire • Mains electricity	6.2.1.2 Electrical charge and current [Electric current as flow of charge only] 6.2.1.3 Current, resistance and potential difference 6.2.1.4 Resistors [Wire as an 'ohmic conductor' not covered; term 'directly proportional' not used; resistance of and applications of thermistors and LDRs not covered; resistance of lamps and diodes covered in Session 30] 6.2.2 Series and parallel circuits [Design and use of series circuits for measurement and testing not covered] 6.2.3.1 Direct and alternating potential difference	• comparing power ratings of appliances in watts (W, kW) • other processes that involve energy transfer: [changing motion, dropping an object,] completing an electrical circuit, [stretching a spring, metabolism of food, burning fuels] • electric current, measured in amperes, in circuits, series and parallel circuits, currents add where branches meet and current as flow of charge • potential difference, measured in volts, battery and bulb ratings; resistance, measured in ohms, as the ratio of potential difference (p.d.) to current • [differences in resistance between conducting and insulating components (quantitative) not covered] • [separation of positive or negative charges when objects are rubbed together: transferof electrons, forces between charged objects not covered] • [the idea of electric field, forces acting across the space between objects not in contact not covered]

Session	Key ideas	AQA GCSE Combined Science specification	KS3 Programme of Study
		[Explanation of difference between direct and alternating potential difference not covered] 6.2.3.2 Mains electricity [Specific potentials of the three wires not covered; dangers not covered] 6.2.4.1 Power	
28. Electromagnetism	• Magnetic poles • Magnetic fields around magnets • Electromagnets • Magnetic fields around electromagnets	6.7.1.1 Poles of a magnet 6.7.1.2 Magnetic fields 6.7.2.1 Electromagnetism	• magnetic poles, attraction and repulsion • magnetic fields by plotting with compass, representation by field lines • Earth's magnetism, compass [and navigation] • the magnetic effect of a current, electromagnets, [D.C. motors (principles only)]
29. Atoms and radioactivity	• The atomic model • Isotopes • Nuclear radiation • Nuclear equations • Radioactive decay	6.4.1.1 The structure of an atom [Radius of atom and of nucleus not covered; change of electron arrangements with absorption or emission of electromagnetic radiation not covered] 6.4.1.2 Mass number, atomic number and isotopes [Number of electrons being equal to number of protons not covered; atoms becoming positive ions if they lose electrons not covered] 6.4.1.3 The development of the model of the atom 6.4.2.1 Radioactive decay and nuclear radiation	• simple (Dalton) atomic model • differences between atoms, elements and compounds • chemical symbols and formulae for elements and compounds

Session	Key ideas	AQA GCSE Combined Science specification	KS3 Programme of Study
		[Range in air and ionising power of alpha, beta, gamma not covered; uses of radiation not covered] 6.4.2.2 Nuclear equations [That gamma emission does not cause mass or charge of nucleus to change not covered] 6.4.2.3 Half-lives and the random nature of radioactive decay 6.4.2.4 Radioactive contamination [Nature of hazard of contamination and comparison with that of irradiation not covered; safety precautions not covered; peer review process not covered]	
30. Physics practicals	• Investigating specific heat capacity • Investigating *IV* characteristics • Using circuit diagrams • Investigating density • Investigating acceleration • Investigating the behaviour of springs • Investigating waves in a ripple tank and on a string • Investigating infrared	Required practical activity 14: Investigation to determine the specific heat capacity of one or more materials Required practical activity 15: Use circuit diagrams to set up and check appropriate circuits to investigate factors affecting the resistance of electrical circuits Required practical activity 16: Use circuit diagrams to construct appropriate circuits to investigate the *I-V* characteristics of circuit elements including a filament lamp, a diode and a resistor at constant temperature Required practical activity 17: Use appropriate apparatus to make and	• forces: associated with deforming objects; stretching and squashing – springs; [with rubbing and friction between surfaces, with pushing things out of the way; resistance to motion of air and water] • forces measured in newtons, measurements of stretch or compression as force is changed • force-extension linear relation; [Hooke's Law as a special case] • pay attention to objectivity and concern for accuracy, precision, repeatability andreproducibility • understand that scientific methods and theories develop as earlier explanations are modified to take account of new evidence and ideas, together with the importance ofpublishing results and peer review

Session	Key ideas	AQA GCSE Combined Science specification	KS3 Programme of Study
	radiation • Physics skills	record measurements to determine the densities of regular and irregular solids [and liquids] Required practical activity 18: Investigate the relationship between force and extension for a spring Required practical activity 19: Investigate the effect of varying the force on the acceleration of an object of constant mass, and of varying the mass of an object on the acceleration produced by a constant force Required practical activity 20: Make observations to identify the suitability of apparatus to measure frequency, wavelength and speed of waves in a ripple tank and waves in a solid and take appropriate measurements Required practical activity 21: Investigate how the amount of infrared radiation absorbed or radiated by a surface depends on the nature of that surface 6.1.1.3 Energy changes in systems 6.2.1.4 Resistors 6.3.1.1 Density of materials 6.3.2.2 Temperature changes in a system and specific heat capacity [Particle model covered in Chemistry Session 12]	• evaluate risks • ask questions and develop a line of enquiry based on observations of the real world,alongside prior knowledge and experience • make predictions using scientific knowledge and understanding • select, plan and carry out the most appropriate types of scientific enquiries to test predictions, including identifying independent, dependent and control variables, whereappropriate • use appropriate techniques, apparatus, and materials during fieldwork and laboratorywork, paying attention to health and safety • make and record observations and measurements using a range of methods for different investigations; and evaluate the reliability of methods and suggest possibleimprovements • apply sampling techniques. • apply mathematical concepts and calculate results • present observations and data using appropriate methods, including tables and graphs • interpret observations and data, including identifying patterns and using observations, measurements and data to draw conclusions • present reasoned explanations, including explaining data in relation to predictions and hypotheses • evaluate data, showing awareness of potential sources of random and systematic error

Session	Key ideas	AQA GCSE Combined Science specification	KS3 Programme of Study
		6.5.3 Forces and elasticity [Interpret data from an investigation of the relationship between force and extension only] 6.6.12 Properties of waves	• identify further questions arising from their results
Physics sections not covered	6.2.1.1, 6.2.4.2, 6.2.4.3, 6.3.1.2 [but this is covered in Chemistry Session 12], 6.3.2.1 [but this is covered in Chemistry Session 12], 6.3.2.3, 6.3.3, 6.5.2, 6.5.4.1.1, 6.5.4.3.4		

SESSION 1: BIOLOGICAL STRUCTURES AND THEIR FUNCTIONS

THINK

Ask the student to read the question for Session 1 in the workbook and scribble down what they know on that topic. Now ask the student to have a go at the questions on Adapt© for Session 1.

Common misconceptions

Students think that plant cells do not have mitochondria.	To address this, always check diagrams and point out the mitochondria as well as the chloroplasts. Reinforce that plants do both photosynthesis and respiration. See key idea card B1.5.
Students think that plant cells do not have a cell membrane.	To address this, always point out plant cell membranes on the inside of the cell wall. The cell wall is for structure; the cell membrane is for controlling what goes in and out of the cell. See key idea cards B1.5 and B1.6.
Students get confused with cell sizes.	To address this, show comparative images of cells with their relative sizes. For example, nerve cells can be very long. See key idea cards B1.2, B1.3, B1.4, B1.5, B1.6, B.17 and B1.8.
Students think that all plant cells have chloroplasts.	To address this, discuss with students the needs of different cells – root hair cells live underground, where there is no light, so they cannot photosynthesise and do not have chloroplasts. See key idea card B1.8.
Students get confused with the relative sizes of molecules, cells, tissues and organs.	To address this, show students comparative images of molecules, cells, tissues and organs with relative sizes. See key idea cards B1.2, B1.3, B2.4, B3.2, B3.4 and B3.6.

CONNECT

The student should watch the video on the powerful idea for the session. The powerful idea for Session 1 is the structure–function relationship.

A key idea in biology is that the structure of a part of a living thing determines its function, or job. This is true for both tiny parts of cells and for organs. The structure can refer to the shape of the cell or organ, or the presence of biological structures, such as mitochondria and chloroplasts.

Teaching tips

Exam questions on organs and cells have a common structure. They will present a cell or organ, and the students will need to know what it does and how it is adapted to fulfilling that function. Many questions can be answered in the same way:

1. Identify the function of the cell or organ.
2. State the adaptation – this could be an organelle (or part of the cell) or a special shape.
3. Link the structure and the function together.

We have included a plant root question in the video. Other examples include palisade cells in the leaf and sperm cells. You can help your students feel confident with any cell or organ by teaching them about a range of functions and a range of adaptations to function. There are questions that allow students to practice applying to a range of cells in the practice questions at the back of the workbook.

For example:

Functions: muscle contractions, nerve impulses, absorption, gas exchange. All of these relate in some way to the 'MRS GREN' life functions.

Structures: presence of mitochodria and chloroplasts, and to a lesser extent ribosomes (higher content); a long shape, such as that found in neurones and root hair cells; connections such as those in neurones.

Practice really does make perfect on this one, so keep coming back to the structure and function of cells and organs you look at in lessons.

SOLVE

Now turn back to the original question with the student. Work through the writing frame in the workbook together. The solutions are given below.

Muscle cells and nerve cells are both involved in movement.

Compare the structure and function of these cells.

Which structures do muscle cells and nerve cells have in common?

- Both cells have a cell membrane, nuclei and cytoplasm.
- Both cells have lots of mitochondria.

How is the shape of each cell different?

- Nerve cells are more branched at the ends than muscle cells.
- Nerve cells are much longer than muscle cells.

Which structures are different in muscle cells and nerve cells?

- Muscle cells have protein filaments.
- Nerve cells have myelin.
- Nerve cells have many connections at both ends.

Muscle cells and nerve cells are both involved in movement.

Compare the structure and function of these cells.

For each of the differences, explain how that structure or shape helps the cell do its job.

- Muscle cells have protein filaments that slide over each other to move the muscle.
- Nerve cells are very long so they can transmit impulses over a long distance.
- Nerve cells have a myelin sheath which acts as an insulator and so makes the transmission of impulses faster.

Now bring together your answers: What do both cells have that they use to make movement happen? What does each cell have that is different that helps it to do its specific job?

- Both cells have mitochondria that release energy – the muscle cell uses the energy to move filaments over each other, and the nerve cell uses the energy to create an electrical impulse.
- Muscle cells are specialised because they have filaments that allow them to shorten, and nerve cells are very long and have a myelin outer layer so they can conduct electrical impulses over a distance quickly.

PRACTISE

Use the additional practice questions on this powerful idea at the back of the workbook in future sessions. Additional practice questions for Session 1 can be found in the Calculate, Explain and Suggest sections.

SESSION 2: MOVING MOLECULES IN LIVING THINGS

THINK

Ask the student to read the question for Session 2 in the Workbook and scribble down what they know on that topic. Now ask the student to have a go at the questions on Adapt© for Session 2.

Common misconceptions

Students need to know that water can move by diffusion.	To address this, show a diagram with water diffusing along a concentration gradient, e.g. when water diffuses through the stomata in leaves (osmosis only happens when there is a partially permeable membrane). See key idea card B2.2.
Students get confused about osmosis and concentration gradients.	To address this, show images of osmosis that show the solute and the water molecules; make sure students realise that water moves from where there is a higher concentration of water (dilute solution) to where there is a lower concentration of water (concentrated solution). See key idea card B2.3.
Students think molecules move into cells because they are 'needed'.	To address this, reinforce that molecules move because of concentration gradients not because they are needed. See key idea cards B2.2, B2.3, B2.4 and B2.5.
Students get confused in osmosis that the solute 'sucks' in the water.	To address this, reinforce that water moves because there is a difference in the concentrations, not because the solute draws in the water. See key idea cards B2.3, B2.5 and B2.6.
Students can get confused over surface area to volume ratio. They think that larger cells would have a higher surface area to volume ratio. Larger cells do have a larger surface, but they have a much bigger volume, so the ratio decreases.	To address this, show students a table with the size of the surface areas, volumes and ratios for different-sized cubes to show the relationship between them.

CONNECT

The student should watch the video on the powerful idea for the session. The powerful idea for Session 2 is moving molecules in living things.

A key idea is that useful substances need to move through multicellular organisms. Molecules move from areas of high concentration to areas of low concentration along a concentration gradient. When water moves across cell membranes it is called osmosis. Shorter distances, a large surface area and larger concentration gradients lead to faster diffusion times.

Teaching tips

Exam questions on moving molecules in living things generally ask students to be able to identify the different ways molecules move and where each one happens in different types of cells, tissues and organs. Students may also be asked to describe factors which would affect the rate of movement and the importance of concentration gradients.

You can help students feel more confident answering questions about moving molecules in living things by showing them that the basic structure of the description of the movement is always the same. In diffusion, molecules move from an area of high concentration to an area of low concentration along the concentration gradient. Osmosis is a special case of this, where molecules of water move from an area of high concentration of water to an area of low concentration of water along the concentration gradient of water and across a semi-permeable membrane.

You can also help students to identify how diffusion is optimised in different parts of living systems, linked to this description: increasing the rate of diffusion relies on making the molecules move faster – by heating them or making a larger surface area for them to diffuse over, or by increasing the concentration gradient. The concentration gradient is increased by increasing the concentration of the molecule that is diffusing, or by increasing how fast the molecule is carried away as it diffuses. Questions may also refer to the osmosis practical; teaching tips for these are included in the practicals section.

Examples of where molecules move include: water moving by osmosis into root hair cells in plants; oxygen and carbon dioxide moving by diffusion in gas exchange from the lungs into the blood and from the blood into body tissues; minerals moving by active transport from the soil into root hair cells in plants, and from the gut into the blood in humans.

SOLVE

Now turn back to the original question with the student. Work through the writing frame in the workbook together. The solutions are given below.

Plants need water for important reactions like photosynthesis.
Explain why plants wilt when they are not watered.

Which plant sub-cellular structures are involved in the structure of the plant?

Plant cell wall, vacuole.

Which plant organs does the water move through from the soil to the leaf? You could list these in order.

Roots → stem → leaves.

How does the water get into the plant? Name and define the process.

Through osmosis in the root hair cells. Osmosis is the diffusion of water from a dilute solution to a concentrated solution through a partially permeable membrane.

How does water move from the roots to the leaves?

Water is pulled up through the xylem vessels in the stem because of transpiration.

Plants need water for important reactions like photosynthesis.
Explain why plants wilt when they are not watered.

Plants lose water from their leaves. How does the water move out of the plant?

Water moves into the leaves by osmosis and is lost from the leaves through the stomata by evaporation from the air spaces.

What happens to the structures in plant cells when they lose too much water?

The vacuole shrinks away from the cell wall making the cell flaccid.

Now bring together what you have written to explain why plants wilt when they do not have enough water.

You will need to include:

- *what water is used for in plant cells*
- *how the water moves in and out of the cells*
- *what happens to the plant cells when the water has gone.*

- Water is needed for the structure of the plant, metabolic processes and photosynthesis.
- Water moves into and out of cells by osmosis because of a difference in solution concentrations. It moves through the plant due to transpiration.
- If plants cells lose too much water, they wilt. This is because the plant cells become flaccid as the vacuole pulls away from the cell wall and the cells cannot replace the water lost in evaporation. Without watering, the plants would eventually die.

PRACTISE

Use the additional practice questions on this powerful idea at the back of the workbook in future sessions. Additional practice questions for Session 2 can be found in the Compare and Describe sections.

SESSION 3: LIFE SYSTEMS

THINK

Ask the student to read the question for Session 3 in the workbook and scribble down what they know on that topic. Now ask the student to have a go at the questions on Adapt© for Session 3.

Common misconceptions

Students think that oxygen and carbon dioxide swap places.	To address this, teach that diffusion needs a concentration gradient, but that the concentration gradient of one does not affect the other. The following sentences should help. Carbon dioxide moves out of the blood into the alveoli down its concentration gradient (because there is more carbon dioxide in the blood than the lungs). Oxygen moves from the alveoli into the blood down its concentration gradient (there is more oxygen in the alveoli than in the blood).
Students think that diffusion happens because cells need molecules.	To address this, ask students why diffusion happens and tackle any misconceptions. Diffusion happens because of concentration gradients. Molecules move down their concentration gradients. Molecules do not diffuse into cells because cells need them. See key idea card B3.2.
Students think that lungs only take in oxygen and only give out carbon dioxide.	To address this, explain that air is taken in and out of the lungs; some of the oxygen is taken in through gas exchange, the rest is breathed out. The carbon dioxide which diffuses out with gas exchange is breathed out with the other carbon dioxide which is naturally in the air.
Students use the word 'respiration' to mean 'breathing'. This can come up in several topics.	To address this, model the use of the correct meaning of the word. Have a discussion about subject specialised words that have one meaning outside of the science classroom and another one in it. See key idea cards B3.2 and B4.2.
Students confuse villi with the folds of the small intestine when looking at images.	To address this, show images of the small intestine at different magnifications to show the increasing number of folds from folds of the small intestine to villi to microvilli. See key idea card B3.6.

CONNECT

The student should watch the video on the powerful idea for the session. The powerful idea for Session 3 is life systems.

A key idea in biology is that molecules can move by diffusion, osmosis or active transport. Molecules move more effectively at specialised exchange surfaces which are adapted for molecules to be transported quickly. Efficient exchange surfaces have large surface areas, a thin membrane, a good blood supply (animals only). Gas exchange systems in animals are also well ventilated

Teaching tips

Exam questions on life systems usually focus on adaptations of different exchange surfaces and the role of enzymes in making larger molecules small enough to be absorbed. Students should know the main adaptations of exchange surfaces: a large surface area and a thin membrane. They should also know specific adaptations for different parts of the body: lungs are well ventilated; the small intestine has villi and microvilli that increase the surface area. They should be able to identify these adaptations in different examples, e.g. the large surface area of a root system in plants. There is a link with Session 2 in terms of how the exchange surface of the cell/tissue/organ helps the movement of molecules.

In the small intestine, enzymes break down the larger molecules from food into molecules that are small enough to be absorbed. Students should be able to explain enzyme action using examples from the small intestine. Questions may be linked with either of the required practicals – investigating the effect of pH on the rate of reaction of amylase enzyme, and using qualitative reagents to test for a range of carbohydrates, lipids and proteins.

To grow students' confidence in answering questions about life systems, give them plenty of practice in identifying the adaptations for diffusion in plants and animals. To understand the idea of enzymes acting as a catalyst, make the link between enzyme-controlled reactions and how catalysts are used in chemistry, paying particular attention to the similarities and differences. Both types of reaction can be speeded up by increasing concentration and increasing surface area, but enzymes will be denatured and the rate of reaction will stop increasing after a specific temperature. Enzymes also have a specific pH they work best at. Students can practice applying this concept by describing and explaining graphs of enzyme action, including identifying the optimum temperature and pH.

SOLVE

Now turn back to the original question with the student. Work through the writing frame in the workbook together. The solutions are given below.

In order to move, animals need energy from respiration.

Compare how the oxygen and glucose needed for respiration enter the body and reach the cells.

For oxygen: Where does oxygen come from? How does it get in? List the parts of the respiratory system that the oxygen passes through. Name the organ where the oxygen meets the blood.

Oxygen comes from the air. It goes through the mouth or nose, then into the lungs. The oxygen travels to the lungs through the trachea, bronchi and into the alveoli. In the alveoli oxygen diffuses into the blood capillaries.

In order to move, animals need energy from respiration.

Compare how the oxygen and glucose needed for respiration enter the body and reach the cells.

Glucose is found in food that needs to be digested before it can be absorbed. Food is broken down by enzymes. What do enzymes do? What do digestive enzymes do?

Enzymes are biological catalysts that speed up chemical reactions. Digestive enzymes help to break down large insoluble molecules into small soluble molecules.

The digestion of glucose happens in the digestive system. List the organs of the digestive system in order, from the mouth to the anus. Name the organ where the glucose meets the blood.

A – mouth, B – gall bladder, C – liver, D – large intestine, E – small intestine, F – stomach

Glucose meets the blood in the small intestine.

Oxygen and glucose both diffuse into the blood. What is diffusion?

Diffusion is the spreading out of the particles of any substance in solution, or particles of a gas, resulting in a net movement from an area of higher concentration to an area of lower concentration.

How are the organs where diffusion happens similar?

- They have a thin layer of cells.
- They have capillary networks.

Bring together your answers in a table: How are the journeys the same? How are they different? You can use tables in the exam for comparison questions.

	Oxygen	Glucose
How substance enters the body	Inhaled in air into the lungs Diffuses into the blood in the alveoli	Ingested in foods containing carbohydrates Moves into the blood in the small intestine
How substance enters cells	Diffuses from the blood into the cells	Diffuses from the blood into cells

PRACTISE

Use the additional practice questions on this powerful idea at the back of the workbook in future sessions. Additional practice questions for Session 3 can be found in the Describe and Explain sections.

SESSION 4: STAYING ALIVE

THINK

Ask the student to read the question for Session 4 in the workbook and scribble down what they know on that topic. Now ask the student to have a go at the questions on Adapt© for Session 4.

Common misconceptions

Students think that arteries always carry oxygenated blood and veins carry deoxygenated blood.	To address this, reinforce that ALL arteries carry blood away from the heart. Veins return blood to the heart (you could link to the Latin *ven* – which means 'to return'). See key idea card B4.5.
Students get confused about which side of the heart does what.	To address this, teach students the following trick to remember which side does which job: • LEFT has four letters, and BODY has four letters • RIGHT has five letters and LUNGS has five letters. See key idea card B4.3.
Students get confused about the sizes of arteries and veins.	To address this, show that the overall sizes of arteries and veins are the same, it is just the size of the lumen that changes. See key idea card B4.5.
Students sometimes think that deoxygenated blood is blue.	To address this, show photos of deoxygenated and oxygenated blood; both are red, but the oxygenated blood is brighter. Veins appear blue because of the way light diffuses through the skin.
Students sometimes think that platelets are cells, but they are actually cell fragments.	To address this, ask students to name the cells in the blood. Reinforce that both red and white blood cells are cells, but platelets are cell fragments. See key idea card B4.4.

CONNECT

The student should watch the video on the powerful idea for the session. The powerful idea for Session 4 is staying alive and how cells, tissues and organs work together.

A key idea in biology is how different parts work together. Cells of similar structure and function group together to form tissues. Tissues with similar functions group together to form organs. Organs are grouped into organ systems to produce an overall function. Organ systems work together to keep organisms alive. Damage to cells, tissues, organs or organ systems call cause disease or death..

Teaching tips

Exam questions on staying alive often refer to the structure and function of the respiratory and circulatory systems. Commonly, questions ask students to label the main structures of the heart and lungs. Parts of the body can be represented in different ways in exam questions, so showing students different diagrams using different viewpoints is a good way to practise this and to make sure students are not confused by slightly different images. Helping them to identify that the wall of the left-hand side of the heart is always thicker helps students to find their way through new images.

It can often help students understand the function of an organ by discussing what happens if that organ is affected by a disease and no longer functions well. For example, if someone has bronchitis then their bronchioles do not function as well, and the person becomes short of breath as their lungs cannot oxygenate their blood as well as the lungs of a healthy person.

Questions asking students to differentiate between arteries, veins and capillaries often come up, and are followed by those asking students to link the structure of the blood vessels to their function. Again, it can help to show students plenty of diagrams and images of cross-sections of the different blood vessels to help identify them and their jobs.

Students also need to be able to recognise the relationship between cells, tissues, organs and organ systems, using the digestive system, respiratory system or circulatory system as examples. It can be helpful for students to learn by heart the route through an organ system to help them make sense of what is happening.

Questions may also refer to lifestyle factors that affect health and increase the risk of coronary heart disease, and may need to interpret data to evaluate the different effects of each factor, such as exercise or a high-fat diet. Give students lots of practise at interpreting different graphs.

SOLVE

Now turn back to the original question with the student. Work through the writing frame in the workbook together. The solutions are given below.

Coronary heart disease is caused by lifestyle factors.

The graph shows the results of a study that looked at the effects of physical activity in men who had heart attacks.

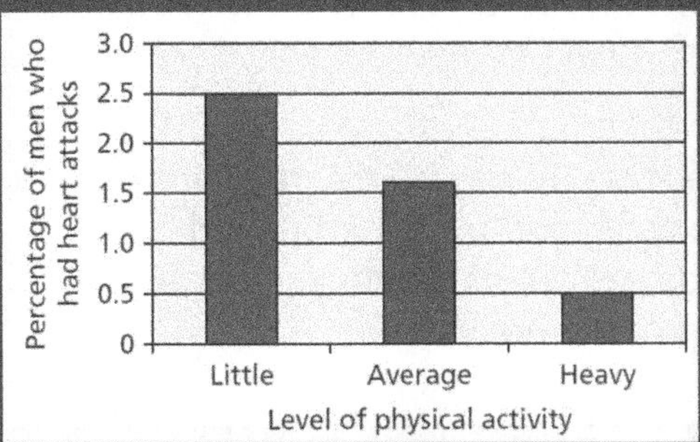

Suggest why people with coronary heart disease get out of breath easily. Use data from the graph to support your answer.

The heart is part of the circulatory system. What are the three types of blood vessel found in the circulatory system?

Arteries, veins and capillaries.

When you exercise your muscle cells need more energy from respiration. What is needed for respiration?

Oxygen and glucose.

Where do the substances needed for respiration come from?

Oxygen comes from the air through the lungs. Glucose comes from the diet through the small intestine.

How does the circulatory system move the substances to the muscles?

Oxygen diffuses from the alveoli into the blood capillaries surrounding them. The heart pumps the blood through the arteries and then the capillaries that surround the muscle.

At the muscle cells oxygen and glucose move from the blood into the muscle cells.

Coronary heart disease is caused by lifestyle factors. Which lifestyle factors can lead to coronary heart disease?

Lifestyle factors include poor diet, lack of exercise and smoking.

What does the data in the chart show you? Describe the results. Remember to include numbers from the graph.

The graph shows that the greater the amount of physical activity the lower the percentage of men who had heart attacks. Men who did little exercise had a 2.5% chance of having a heart attack, but men who did heavy exercise only had a 0.5% chance.

What happens to your breathing and heart rate when you exercise?

Both breathing and heart rate increase during exercise. Breathing rate increases the rate of gas exchange and heart rate increases the rate of blood reaching the cells.

Why does this happen?

Increased exercise leads to more muscle contraction which needs more energy from respiration. The increased oxygen needed for respiration comes from a faster ventilation rate. This also helps the carbon dioxide made by respiration leave the body. Respiration also needs glucose, and the increased heart rate allows the blood carrying oxygen and glucose to reach cells faster.

Now bring together what you have written about how the heart works and what happens when you exercise to answer the question: Suggest why people with coronary heart disease get out of breath more easily.

With coronary heart disease the blood vessels around the heart become blocked. This means the heart cannot pump as well, so less oxygen and nutrients get to the cells, and more carbon dioxide builds up. To make up for this your breathing rate increases and you get out of breath as your body tries to get more oxygen into the body and more carbon dioxide out of the body.

PRACTISE

Use the additional practice questions on this powerful idea at the back of the workbook in future sessions. Additional practice questions for Session 4 can be found in the Compare and Evaluate sections.

SESSION 5: NERVOUS CONTROL

THINK

Ask the student to read the question for Session 5 in the workbook and scribble down what they know on that topic. Now ask the student to have a go at the questions on Adapt© for Session 5.

Common misconceptions

Students often find it difficult to understand that the spinal cord can make decisions in reflex reactions.	To address this, reinforce that both the brain and the spinal cord are part of the central nervous system and both can coordinate responses. See key idea cards B5.1, B5.2 and B5.4.
Students can find it difficult to visualise a neurone and how it relates to a nerve.	To address this, show an electron micrograph of a nerve which shows the individual nerve cells (neurones).
Students may think that changes in the environment are just external changes, but many of the body responses are due to internal changes within the body.	To address this, discuss nerve responses that occur due to internal changes, like an increased heart rate during exercise.
Students may get confused between the terminology for receptors and effectors.	To address this, ask students to label a reflex arc, make links between receptor and reception, effector and effective. See key idea card B5.3.
Students may find it difficult to visualise how long nerve cells can be.	To address this, show students some facts about the longest nerves in various animals (in humans, the sciatic nerve is as long as the distance between the spine and the foot). See key idea cards B5.4 and B5.5.

CONNECT

The student should watch the video on the powerful idea for the session. The powerful idea for Session 5 is nervous control.

A key idea in biology is that our bodies have control systems that allow for communication between body parts that leads to changes to keep our internal environment stable. In nervous control, changes are detected by receptors which send impulses to the coordination centre (brain or spinal cord) via electrical signals in nerves. The response is sent via nerves to effectors which bring about a response. Nervous response is fast and short acting.

Teaching tips

Exam questions on nervous control usually focus on either the structure of the cells and tissues of the nervous system or knowledge of the reflex arc. Linking the structure of nerve cells to their function often comes up in questions. Students should have plenty of practice in identifying the main cell parts on diagrams of neurones, and linking them to the specific adaptations that neurones have.

Students should be familiar with a range of reflex actions and be able to explain how they are brought about using the reflex arc. Commonly, questions ask students to identify different parts of the reflex arc on a diagram. Give students plenty of practice at applying what they know to different diagrams and asking them to describe different parts of the process.

Questions may also refer to the required practical planning and carrying out an investigation into the effect of a factor on human reaction time. This is a good opportunity for students to practise plotting and interpreting graphs, as well as discussing what makes a good experiment, identifying variables and analysing results.

You can support students to grow their confidence in this area by getting them to learn the reflex arc by heart, making sure they get the motor neurone and sensory neurone in the right order.

SOLVE

Now turn back to the original question with the student. Work through the writing frame in the workbook together. The solutions are given below.

Humans have reflexes to protect them from harm.

Show how a reflex arc makes you quickly move your hand away from something hot.

Which organs are in the central nervous system?

Brain and spinal cord.

Neurones or nerve cells are specialised to conduct nerve impulses. How are neurones specialised?

Neurones are very long, which allows cells to connect the central nervous system to all parts of the body. There are fine branches which allow neurones to connect with other cells. Nerves are insulated by a myelin sheath so they can transmit information quickly.

What are the three types of neurone found in a reflex arc?

Sensory neurone, relay neurone, motor neurone.

Draw the reflex arc and list the steps. Make sure you include the names of each type of neurone.

See key idea card B5.4.

Now bring together all the steps in order and link the fast reflex to the need to move quickly away from the heat.

- *When the hand touches something hot it needs to move quickly because…*

  The heat can burn and damage the cells.

- *The hand moves quickly because this is a reflex action. The steps of a reflex action are...*

  The stimulus is the hotplate. When the hand touches it the skin receptors detect the change in temperature and send an impulse to the spinal cord (central nervous system) through the sensory neurones. The spinal cord coordinates the response and sends an impulse through the motor neurones to the effectors, which are the muscles in the arm and hand. The response is to move the hand away.

What is another example of a reflex action?

Humans have reflexes to protect them from harm.   **Show how a reflex arc makes you quickly move your hand away from something hot.**
Blinking when something comes close to our eyes.

PRACTISE

Use the additional practice questions on this powerful idea at the back of the workbook in future sessions. Additional practice questions for Session 5 can be found in the Compare and Show sections.

SESSION 6: HORMONAL CONTROL

THINK

Ask the student to read the question for Session 6 in the workbook and scribble down what they know on that topic. Now ask the student to have a go at the questions on Adapt© for Session 6.

Common misconceptions

Students may think that the point of homeostasis is to reach a static point, but it is naturally always changing.	To address this, use blood glucose concentrations as an example, where the body has to constantly monitor and correct, which is why people with Type 1 diabetes have to test their blood sugar level throughout the day. See key idea card B6.4.
Students may think that changes in the environment are just external changes, but many of the body responses are due to internal changes within the body.	To address this, talk about changes in water balance that affect internal cells; using plant cells as an example, show how the vacuole stores and releases water depending on how much water is in the cytoplasm. See key idea cards B6.1 and B6.2.
Students often think that Type 1 diabetes is genetic.	To address this, point out that neither Type 1 nor Type 2 diabetes is a genetic disease, but certain genes may increase the risk of getting diabetes. See key idea card B6.4.
Students may think that testosterone is only made by men, and that oestrogen, progesterone, FSH and LH are only made by women, but both sexes make all the hormones.	To address this, reinforce that all sexes make all of the sex hormones, but in different amounts.
Students may think of being 'hormonal' or having hormones as feeling emotional.	To address this, discuss the various effects of hormones on the body, and that while they can affect our emotions they also have many other roles. See key idea cards B6.3, B6.4, B6.5 and B6.6.

CONNECT

The student should watch the video on the powerful idea for the session. The powerful idea for Session 6 is hormonal control.

A key idea in biology is that of communication between parts of the body to allow a stable internal environment. Communication can be by hormones. In hormonal control receptors sense when a hormone need to be produced, hormones are then secreted by glands and transported to effector organs where they have an effect. Hormonal control is slower than nerves, but their effects last longer.

Teaching tips

Exam questions on hormonal control commonly ask students to:

- identify the main glands of the endocrine system on a diagram
- explain blood sugar control by the pancreas and insulin
- interpret graphs of blood sugar in diabetes
- compare hormonal control to nervous control
- describe homeostasis as a process of keeping the internal environment the same
- name the hormones involved in puberty and the menstrual cycle, and describe what each one does
- evaluate hormonal and non-hormonal methods of contraception.

Teaching about insulin and blood glucose is a good opportunity for interpreting graphs and comparing blood glucose in people with diabetes and those without; use this as a way for students to explain the effects of insulin.

Questions on reproductive hormones are often linked with the role of hormones in contraceptives. Give students plenty of practice in evaluating contraceptives, to include reasons – other than scientific ones – why couples might prefer different methods of contraception for cultural, ethical or financial reasons.

With homeostasis, make sure students can describe the different responses to changing levels, and give them practise at linking different changes to the body's response.

SOLVE

Now turn back to the original question with the student. Work through the writing frame in the workbook together. The solutions are given below.

The graph shows how the amount of glucose in a healthy person's bloodstream changes after eating a meal.

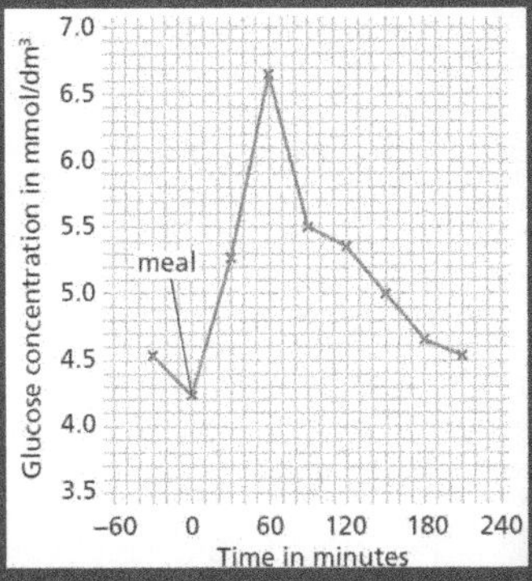

a. Explain how your body controls blood sugar, by homeostasis, using hormones.

b. Diabetes is a condition in which blood sugar is uncontrolled. Explain how the shape of the graph would look different in someone with diabetes.

For part (a): Where does the sugar in the blood come from?

Glucose (sugar) comes from carbohydrates in the diet.

For part (a): How is glucose (sugar) removed from the blood?

Glucose is stored as glycogen in the liver.

For part (a): How does the body know if there is too much glucose in the blood?

The pancreas detects sugar levels.

For part (a): Now put it all together to explain how your body controls blood sugar.

After a meal your blood glucose increases. The pancreas detects the increase in glucose and releases insulin. Insulin acts on your cells to take in glucose and store it as glycogen and use it in cell respiration. This leads to a drop in blood glucose.

For part (b): Look at the graph. Describe the shape of the graph in your own words.

The line goes down to start with then shoots up quickly after the meal. After the peak, it drops back down again.

For part (b): People who have a poor diet may develop Type 2 diabetes. What happens to blood sugar levels in someone with diabetes?

In Type 2 diabetes the pancreas will not make enough insulin and/or the cells do not respond to it. So blood sugar levels will remain high after someone with diabetes has eaten a meal.

For part (b): Draw a sketch on the graph in the question to show what would happen to blood sugar in someone with diabetes. Remember that your sketch is a good guess or estimate of how the line would look. How is your sketch different from the graph for the healthy person?

Your sketch should start slightly higher and then increase by the same amount up to 60 minutes but then should continue rising and stay higher for the rest of the graph.

PRACTISE

Use the additional practice questions on this powerful idea at the back of the workbook in future sessions. Additional practice questions for Session 6 can be found in the Compare and Define sections.

SESSION 7: DISEASE AND IMMUNITY

THINK

Ask the student to read the question for Session 7 in the workbook and scribble down what they know on that topic. Now ask the student to have a go at the questions on Adapt© for Session 7.

Common misconceptions

Students may not realise that antibodies are not alive, but are small proteins made by lymphocytes.	To address this, ask students if they think antibodies are alive, reinforce that they are proteins which are made by living cells, but are not alive. See key idea card B7.4.
Students may think that antibiotics kill viruses.	To address this, discuss that some chemicals like bleach kill all cells (including us) but other chemicals like antibiotics, only kill bacteria. See key idea card B7.7.
Students may have numerous misconceptions about vaccination.	To address this, use websites directly addressing current antivaccination theories and allow students to ask questions. See key idea card B7.7.
Students may have numerous misconceptions about how diseases like HIV are spread.	To address this, use websites for HIV education that directly combat common misconceptions. See key idea card B7.6.
Students may not realise that pathogens can also infect plants as well as animals.	To address this, show images of plants infected with the rose black fungus and tobacco mosaic virus. See key idea card B7.6.

CONNECT

The student should watch the video on the powerful idea for the session. The powerful idea for Session 7 is disease and immunity

A key idea in biology is the role of the immune system in fighting pathogenic microbes and the toxins they produce. The immune system is made of white blood cells which identify pathogenic proteins as foreign and then attache to and destroy them. Protection remains as some of the matching white blood cells remain so that if the same pathogen enters the body the white blood cells can quickly multiply to prevent disease.

Teaching tips

Exam questions on disease and immunity require students to be able to remember several types of pathogen and the plant and animal diseases they cause. Questions also ask students to describe the barriers the body has to pathogens and the basic immune response. There are often data questions related to immunity, so students need to be confident in reading data from a graph and describing what is happening.

To help students remember the required detail, they could make and play a mix and match card game with pathogens and symptoms. Asking students to label, on a diagram, areas of the body that have barriers to infection may also be helpful.

Students should be able to describe the role of white blood cells in fighting any pathogens that get past the barriers. Make sure students can differentiate between an antibody and an antigen, and put the steps of the immune response in order. You can help students to grow their confidence by practising describing the immune response to a number of triggers, such as vaccines and pathogens encountered naturally.

Use graphs to show the difference between the immune response the first time a pathogen is encountered and the second time, and then support students to explain how this relates to the success of vaccines.

For treatments, make sure students understand that antibiotics can only be used for bacterial infections. Questions on the use of antibiotics are often linked to the overuse of antibiotics and subsequent antibiotic resistance.

SOLVE

Now turn back to the original question with the student. Work through the writing frame in the workbook together. The solutions are given below.

Your immune system stops infections caused by pathogens (harmful microbes).

Explain how vaccines trigger this process and how this protects you.

What is the general term for an organism or virus that causes infectious diseases? What kind of organisms other than viruses can cause infectious diseases?

Infectious diseases are caused by pathogens. Pathogens are organisms like bacteria, fungi or protists that cause a disease.

What are the barriers that stop pathogens getting into the body?

Skin, nose hair and mucus, and stomach acid.

If pathogens do get in, how does the body recognise them?

Our body recognises any cell that is not a human cell. (This is because of the non-human proteins on their outer membranes.)

Which cells are involved in fighting the pathogens?

White blood cells – phagocytes and lymphocytes.

What do these cells do?

Phagocytes ingest many types of pathogen. Lymphocytes make antibodies which stick pathogens together, and antitoxins which neutralise toxins.

Some of the white blood cells that have fought off an infection remain in the body. How does this help you if you are infected with the same pathogen again?

Once you have immunity, if you have an infection with the same pathogen your response will be faster and bigger.

Why does this only work for that specific pathogen?

Immunity comes from the lymphocytes. Lymphocytes are specific to a single type of pathogen.

Vaccines contain a tiny part of a pathogen, or a pathogen that has been damaged so it does not cause disease. How does a vaccine stop you getting ill?

When your body is infected with a tiny part of the pathogen your immune response is triggered. Specific lymphocytes for the pathogen are activated and they produce the specific antibodies needed to fight the pathogen. If you are infected with the real pathogen the secondary immune response is faster and bigger. The antibodies then stick to the pathogen and stop it from giving you the disease.

PRACTISE

Use the additional practice questions on this powerful idea at the back of the workbook in future sessions. Additional practice questions for Session 7 can be found in the Define and Explain sections.

SESSION 8: LIFE CYCLES

THINK

Ask the student to read the question for Session 8 in the workbook and scribble down what they know on that topic. Now ask the student to have a go at the questions on Adapt© for Session 8.

Common misconceptions

Students often think that plants do not do cell respiration.	To address this, show an image of a plant cell with chloroplasts and mitochondria; reinforce that plants photosynthesise to make sugar, and respire to release the energy from sugar. See key idea card B8.3.
Students often think that plants do not need oxygen.	To address this, similar to above, make sure students know that plants also need to respire, so they use the waste oxygen produced by photosynthesis to do respiration. See key idea card B8.3.
Students often think that a plant's mass comes from the soil.	To address this, show students a plant and ask them where the mass has come from, reinforce that the carbon in the plant comes from fixed gaseous carbon dioxide from the atmosphere, which becomes the molecules like cellulose and starch in the plant cells. See key idea cards B8.2 and B8.3.
Students often describe respiration as 'making' energy. Energy cannot be made, but is released.	To address this, talk about the importance of clear communication in science, and reinforce that energy is released from sugar in respiration, it cannot be made by respiration. See key idea card B8.7.
Students often draw the arrows in food chains from the thing eating to the thing being eaten. The arrows show energy flow, so always go from the thing being eaten to the thing doing the eating.	To address this, show a food chain without arrows, ask students to predict which way the arrows should go, and then identify the correct way. See key idea card B8.6.

CONNECT

The student should watch the video on the powerful idea for the session. The powerful idea for Session 8 is life cycles.

A key idea in biology is that substances that make up living things are recycled. The carbon dioxide made during cell respiration is used by plants in photosynthesis to make sugar. This sugar in turn makes up plant bodies which animals eat to make their own bodies. When organisms die bodies are decomposed by microorganisms to be resused.

Teaching tips

Exam questions on life cycles include nutrient cycles, food chains and the interconnection between respiration and photosynthesis. There are a lot of sub-topics that are all related to the powerful idea of energy flow in living systems. The reactions of photosynthesis and respiration underpin all the other content, so it is crucial that students understand the two reactions so that they can apply them to:

- interpret predator–prey cycles
- interpret or explain diagrams of the water cycle or carbon cycle
- explain what happens to energy and matter in food chains.

Questions on life cycles are frequently linked to global warming and the interpretation and plotting of graphs, so make sure students get plenty of practice at analysing different sources of data linked to global warming.

Questions on food chains often require students to identify the different roles of organisms in an ecosystem. The arrow in a food chain is the key to understanding what is being communicated – the arrow represents the flow of energy.

You can help students to grow their confidence in this topic by practising identifying the producers and consumers in lots of different food chains, and reminding them that the energy always comes from the Sun and is fixed by plants.

Students are required to know and be able to identify the reactants and products in the word equations for respiration and photosynthesis. Photosynthesis questions are often linked to the required practical. Make sure students are familiar with the different types of apparatus used to measure photosynthesis and the types of graph that would be produced.

SOLVE

Now turn back to the original question with the student. Work through the writing frame in the workbook together. The solutions are given below.

The materials that living things are made of have existed on Earth for millions of years. Describe how carbon moves through living things via respiration, photosynthesis and decomposition. Include the conditions needed for each process.

Plants are the first user of carbon, so we will start there. How do plants use carbon? What do they make it into? Write an equation for the process.

Plants take in carbon dioxide and convert it into glucose in photosynthesis.

$$\text{carbon dioxide + water} \xrightarrow{\text{Light and chlorophyll}} \text{glucose + oxygen}$$

The glucose can be:

- used in cell respiration
- converted into other molecules to be used in growth and repair
- converted into cellulose for cell walls
- stored as starch.

Carbon can be passed on by the plant dying or being eaten. What happens to a plant if it dies? What are the conditions for decomposition?

When a plant dies it will decompose. Bacteria and other decomposers in the soil will recycle the plant back into nutrients, which can be taken in by other organisms.

Carbon is passed along the food chain. Draw and label a food chain, naming each level. Use the labels: producer, primary consumer, secondary consumer, tertiary consumer.

For an example food chain, see key idea card B8.6.

Glucose contains carbon. Carbon in glucose is used in the process of respiration to release energy in plants and animals. Oxygen is needed, and carbon dioxide and water are produced. Show this as a word equation.

oxygen + glucose → carbon dioxide + water + energy

Carbon can be released from fossil fuels when they are burned. Now sum up how carbon gets into organisms, and how it leaves. You could draw a labelled diagram.

For an example diagram of the carbon cycle, see key idea card B8.4.

PRACTISE

Use the additional practice questions on this powerful idea at the back of the workbook in future sessions. Additional practice questions for Session 8 can be found in the Describe and Suggest sections.

SESSION 9: REPRODUCTION AND INHERITANCE

THINK

Ask the student to read the question for Session 9 in the workbook and scribble down what they know on that topic. Now ask the student to have a go at the questions on Adapt© for Session 9.

Common misconceptions

Students often think that the processes that bring about variation (mutation, meiosis, sexual reproduction) produce 'useful' variants, and that species are trying to adapt.	To address this, discuss that the variation produced is random, but natural selection selects the variants that are most likely to produce more offspring. If a useful variant never occurs, then it can never be selected. See key idea cards B9.3 and B9.4.
Students often think that natural selection and evolution are the same thing.	To address this, ask students to differentiate between the two – give clear definitions: evolution is the change in a species over time; natural selection is one of the ways that this can happen. See key idea cards B9.3 and B9.4.
Students often find it difficult to understand the time organisms have been evolving for.	To address this, use the analogy of a 24-hour clock. Tell students that if the Earth has been around for a day, the following times would show the evolution of humans in relation to the first cells: • 4 a.m. – first cells • 8 p.m. – first multicellular organisms • 11 p.m. – first mammals • 11:59 p.m. – first humans. See key idea cards B9.3 and B9.4.
Students often think humans evolved from monkeys.	To address this, show an evolutionary tree. Monkeys and humans both evolved from a common ancestor, but since the pathway split, monkeys have been evolving at the same time as humans have, so we have both changed from our common ancestor.
Students often think that the peppered moth 'turned' black.	To address this, reinforce that moth populations show variation. Show populations of moths with variation. Ask students which moths would be the most successful in both polluted and clean areas, and ask them to predict what would happen over time. See key idea card B9.4.

CONNECT

The student should watch the video on the powerful idea for the session. The powerful idea for Session 9 is reproduction and inheritance.

A key idea in biology is that characteristics are coded for in DNA. When plants and animal reproduce they produce gametes with a random half of the DNA. DNA can also change through mutation. These processes lead to variation between individuals, and some of those differences help individual survive.

Teaching tips

Exam questions on reproduction and inheritance refer to meiosis and other sources of variation that allow evolution by natural selection.

Questions commonly ask students to:

- complete Punnett square diagrams to show the inheritance of sex
- draw or complete diagrams to show how chromosomes behave during meiosis
- explain why meiosis leads to variation in offspring
- compare asexual and sexual reproduction.

When teaching natural selection, model an explanation with one example, like the melanic moths, then ask students to use that model to explain other examples of natural selection like the Galapagos finch beaks. Build students' confidence by teaching them the following steps, which can be applied to different situations where an organism has changed over time:

1. There is variation in a species.
2. Some variations allow organisms to survive better when there is a change in the environment.
3. Members of the species with the beneficial allele will survive better.
4. These members of the species will reproduce and pass on the beneficial allele, so the characteristic becomes more common over time.

Give students practise at describing the process of meiosis and listing other sources of variation. Ask them to use Punnett squares to explain why every time a couple has a child there is a 50:50 chance of having either a boy or a girl.

This topic requires lots of specific vocabulary, so it would be useful to practise this with the students using questions on Adapt©.

SOLVE

Now turn back to the original question with the student. Work through the writing frame in the workbook together. The solutions are given below.

The organisms alive today are very different from the organisms that lived millions of years ago. Peppered moths are either white and speckled, or black. During the daytime peppered moths rest on branches of trees. In the 20th century the bark of these trees became covered in black soot.

Show how peppered moths got darker in colour through natural selection when the trees they lived on became sooty.

In the original population of moths, most moths were light coloured, and a few were dark. Describe why some organisms look different.

Variation in a population is caused by mutation, meiosis and sexual reproduction. Variation is why some organisms look different.

When the trees got sooty, more of the dark-coloured moths survived. Why?

Dark-coloured moths were more likely to survive because it was easier for them to hide from predators on the dark sooty trees.

Camouflage like this is an adaptation to not being eaten. Animals and plants have many adaptations that help them survive. Write or draw four examples below.

There are many different adaptations. Some examples include:

- Polar bears have thick fur to insulate them from the cold.
- Woodpeckers have long tongues to take insects from crevices in trees.
- Cheetahs run fast to catch prey.
- Tigers have stripes to camouflage them from prey.
- Flowers have bright colours to attract pollinators.

More of the next generation of moths were darker coloured. Why?

More dark moths were able to avoid being eaten by predators, so they were able to reproduce and pass on the dark colouring to more offspring.

Over time the whole population became darker coloured. Bring together the steps that led to this.

- Environmental change – trees became darker because of sooty pollution.
- Moths were a variety of different colours because of mutation, meiosis and sexual reproduction.
- Moths that were darker were better able to hide from predators.
- The darker moths survived and reproduced, having offspring which were also dark.
- Over the generations the population of moths grew darker.

PRACTISE

Use the additional practice questions on this powerful idea at the back of the workbook in future sessions. An additional practice questions for Session 9 can be found in the Show section.

SESSION 10: GENETICS AND GENETIC ENGINEERING

THINK

Ask the student to read the question for Session 10 in the workbook and scribble down what they know on that topic. Now ask the student to have a go at the questions on Adapt© for Session 10.

Common misconceptions

Students often think that dominant alleles are 'better'.	To address this, reinforce the idea that dominant traits just mean always expressed – dominant alleles are expressed in a heterozygous and homozygous genotype. Recessive alleles are not weaker, you just need two copies of the allele for a recessive trait to be expressed. You could model this with a neutral trait like ear attachment – dominant ears (unattached lobes) are not 'better' than recessive ears (attached lobes). See key idea card B10.4.
Students often think that dominant alleles will be more common.	To address this, discuss recessive traits in a population: In Finland the majority of people will have blond hair, despite it being a recessive trait. See key idea card B10.4.
Students may get confused with the difference between a chromosome, a gene and an allele.	To address this, you could use the analogy of a recipe book: the chromosome is the recipe book, the gene is the type of recipe (e.g. brownie recipe), the allele is the version of the recipe (e.g. gluten-free brownie, dark chocolate brownie). See key idea card B10.3.
Students may think that all traits can be genetic.	To address this, discuss the fact that many traits are also linked to the environment, so skin colour is partly genetic and partly due to how much sunbathing someone has done. See key idea card B10.1.
Students may have heard scare stories about genetic engineering.	To address this, asks students to list their fears, or what they have heard about genetic engineering and use websites that directly tackle the common misconceptions. See key idea card B10.5.

CONNECT

The student should watch the video on the powerful idea for the session. The powerful idea for Session 10 is genetics and genetic engineering

A key idea in biology is that the code in the DNA, in eukaryotes this is found in the nucleus. Lengths of DNA that code for a characteristic are called genes, and individuals have different sets of these called alleles. Exact copies of cells are made through mitosis. Humans use genetic engineering to make products, plants and animals for food and medicine.

Teaching tips

Exam questions on genetics and genetic engineering usually focus on Punnett squares, explaining and evaluating genetic engineering, and understanding stem cells and mitosis. Students will need to be confident in working out basic probabilities from Punnett squares, and need to understand the difference between cell division for growth and cell division for sexual reproduction.

This is a vocabulary-dense topic, and students could make language games to help them learn the definitions. For genetics, students should be given lots of practice of completing Punnett squares and calculating probabilities, especially with the named examples of cystic fibrosis and polydactyly.

For genetic engineering, students should be able to describe the similarities and differences between selective breeding and genetic engineering, and use examples to show when each would be used.

Students need to be familiar with the advantages and disadvantages of sexual and asexual reproduction in commercial examples. You can help students grow their confidence in this area by looking at a range of examples where organisms are produced and what the advantages and disadvantages of the technique are. The main idea is that asexual reproduction is quick, cheap and profitable, as many thousands of organisms (mainly plants) can be made that are reliably identical to the parent. Sexual reproduction takes longer and the resulting genotypes are more variable, but the advantage of this is that if a disease is encountered then the sexually reproduced organisms will be more resistant as they have more variability in their genotype.

For stem cells, make sure students know that these cells are not linked to genetic engineering as the DNA is not changed. Students should be able to explain the importance of stem cells being undifferentiated, and how that means they have the potential to become any type of cell. Give students the chance to describe mitosis and how the amount of DNA changes through the process.

SOLVE

Now turn back to the original question with the student. Work through the writing frame in the workbook together. The solutions are given below.

Polydactyly is an inherited condition caused by a dominant allele.

A pregnant woman does not have polydactyly, but her husband Albert does.

Albert's mother does not have polydactyly.

Determine the possible genotypes and phenotypes for their child.

Complete the Punnett square shown to work out your answer.

Use D for the dominant allele and d for the recessive allele.

		Mother (dd) gametes	
		d	d
Father (Dd) gametes	D		
	d		

To help answer the question, let us recap some genetic terms. What is a dominant trait?

A dominant trait is one where only one allele of a gene needs to be present for the trait to be expressed.

What is the difference between a dominant and a recessive trait?

A dominant allele is always expressed if it is present in the genotype.

Two copies of a recessive allele need to be present for the recessive trait to be expressed.

What is the difference between genotype and phenotype?

Genotype shows the alleles present for a gene – the genetic make-up. Phenotype is how the gene is expressed or shown in the individual.

In the question you are told to use D for the dominant allele and d for the recessive allele. Write down the genotypes of someone with polydactyly. There are two possible genotypes.

DD or Dd.

Now have a go at completing the Punnett square in the question. In each of the four empty squares, you need to write the letters of two alleles. One must come from the mother and one from the father.

See the Punnett square in the solutions section of the workbook.

From the genotypes you have written for the offspring, identify which genotype is heterozygous, and which is homozygous.

The heterozygous genotype is: Dd

The homozygous genotype is: dd

Now identify the phenotype of each combination underneath the alleles in the Punnett square. In each of the four squares you should write 'polydactyly' or 'unaffected'.

		Mother (dd) gametes	
		d	d
Father (Dd) gametes	D	Dd polydactyly	Dd polydactyly
	d	dd unaffected	dd unaffected

Complete the sentences below to describe the probability of the child having polydactyly. You can write probability as a fraction, percentage or ratio.

The probability of the child having polydactyly is: 0.5, 1/2 or 50%

The probability of the child **not** having polydactyly is: 0.5, 1/2 or 50%.

PRACTISE

Use the additional practice questions on this powerful idea at the back of the workbook in future sessions. Additional practice questions for Session 10 can be found in the Compare and Determine sections.

SESSION 11: BIOLOGY PRACTICALS

THINK

Ask the student to read the question for Session 11 in the workbook and scribble down what they know on that topic. Now ask the student to have a go at the questions on Adapt© for Session 11.

Common misconceptions

Students may get confused between independent and dependent variables.	To address this, get students to practise identifying both types of variable from a variety of biology practicals.
Students often start experiments with preconceived ideas over what they will see, which can impact what they observe.	To address this, as a group, ask students to predict what they will see, and try to get a range of ideas to show that students have different ideas. Then discuss the observations after data collection.
Students can often be confused about measurements, particularly relative sizes and magnification.	To address this, show images of different biological objects from molecules, to cells, to ecosystems, and include their sizes (with SI units) so that students are better able to visualise the relative sizes. For practice, ask students to give the sizes of the objects using different units (e.g. change from mm to m, or km to m).
Students often confuse accuracy and precision.	To address this, a useful image is that of a bullseye: an accurate measurement is close to the real value, so all the darts are in the bullseye, but not necessarily close together; a precise measurement is one that is clustered together, so the darts are close to each other, but not necessarily in the bullseye. An accurate and precise measurement would be darts close together and in the bullseye.
Students often confuse describing data and explaining data.	To address this, ask students to give answers in a table: in one column have a description of part of the data (e.g. positive correlation trend line); in the second column have the explanation in terms of the data (e.g. as the environment gets warmer people sweat more).

CONNECT

The student should watch the video on the powerful idea for the session. The powerful idea for Session 11 is biology practicals. A key idea in biology is how knowledge has developed. The development of the microscope allowed the identification of the cell, for example, while other experiments show what living things are made of and how reactions occur.

Teaching tips

Exam questions about the required practicals are used as mechanisms to ask questions about the range of practical skills questions outlined in the specification under 'Working scientifically – Experimental skills and investigation'.

Questions involving microscopes require students to describe the process of using a microscope and being able to use an equation to calculate the actual sizes of specimens. Give students plenty of practice at calculating sizes, as they often find this difficult.

Questions involving osmosis practicals require students to be able to understand different ways of measuring osmosis in plant tissues. To gain confidence in unfamiliar experiments, show students a variety of experiments that require this, and ask them to identify the key variables. Students also need to practise interpreting graphs and plotting the results of these experiments.

Food tests require students to memorise different methodologies, so a mix and match game might make this more fun. Show students different foods and ask them to predict what type of molecule might be found, and then ask them to explain how they would find out. This is a good chance for students to identify risks and to write risk assessments for apparatus and chemicals, such as Bunsen burners and the different reagents needed.

The experiment on the effect of pH on the enzyme amylase is a good opportunity for students to practise choosing appropriate apparatus. Give students graphs to interpret, and get them to practise explaining the results using what they know about enzymes.

Investigating the effect of light on the rate of photosynthesis is a good opportunity to allow students to pick different ways of measuring the dependent variable. Show students different experimental set-ups and get them to identify controls and the dependent variable. Use graphs of the data collected to allow students to describe what they know about photosynthesis.

Reaction time experiments are useful for discussing the role of ethics in experimentation; you can discuss what is and is not ethical, and how this might limit the data that can be collected. Graphs can be used to practise interpretation, and to get students to link what they know about the reflex arc with the data.

Population sampling is a good opportunity for students to practise writing hypotheses and considering the ethics of sampling from the environment. Mathematical data can be given to students for them to practise calculating different types of average and plotting graphs. It is also a useful experiment to discuss the limitations of the scientific method in getting a representative sample.

SOLVE

Now turn back to the original question with the student. Work through the writing frame in the workbook together. The solutions are given below.

A student carried out an experiment into the effect of light intensity on the rate of photosynthesis.

The student's results are shown in the table.

Distance of lamp from pondweed (cm)	Number of bubbles per minute		
	Test 1	Test 2	Test 3
10	99	121	124
15	41	50	54
20	20	30	32
25	11	17	16
30	8	13	14

Describe how the student could analyse their results.

What equipment would the student have used to measure the rate of photosynthesis?

Equipment – beaker, funnel, upturned measuring cylinder/test tube, aquatic plant.

To measure rate you need to measure how fast a product is made or how fast a reactant is used. Did the student measure a reactant or product in this investigation?

You would measure the rate of oxygen production (the product), by reading the volume collected in the measuring cylinder at regular time intervals.

How did they change the light intensity?

They changed the distance of the lamp from the plant.

Why does the rate of photosynthesis change when light intensity changes?

As light intensity increases more light can be captured by the chlorophyll in the leaves which increases the rate of photosynthesis. This means that more oxygen is produced as a (waste) product.

From the results shown you can see the student carried out three repeats (tests) of the experiment. What calculation could you do to get a representative result for the bubbles counted at each distance?

To get a representative result, calculate the mean (average).

Is the data repeatable or are there any anomalous results?

The results for test 2 and test 3 look repeatable, but test 1 results are anomalous – they do not agree. This could be because the student did not start counting the bubbles as soon as the timer was started. Or perhaps their method for counting the bubbles improved after the first test. The test 1 results would need to be discounted from the calculated average or the test repeated a fourth time.

What kind of graph could they plot? What shape would the graph be?

Graph – a line graph with distance of lamp from pondweed (cm) along the *x*-axis and number of bubbles produced per minute on the *y*-axis. The sketch should show a downwards slope from left to right with a steeper slope to start with, levelling off towards 30 cm.

PRACTISE

Use the additional practice questions on this powerful idea at the back of the workbook in future sessions. Additional practice questions for Session 11 can be found in the Design, Plan, Predict, Sketch and Suggest sections.

SESSION 12: THE PARTICLE MODEL

THINK

Ask the student to read the question for Session 12 in the workbook and scribble down what they know on that topic. Now ask the student to have a go at the questions on Adapt© for Session 12.

Common misconceptions

Students think that particles of a solid only move when the object is moving or is moved.	To address this, explain that scientists use models to explain what is happening on a submicroscopic level and according to the particle model. All particles are always moving, but we cannot see them. See key idea card C12.2.
Students think that particles of a gas are not in constant motion.	To address this, spray some perfume or air freshener in the air and ask the students to put their hand up when they smell it. See key idea card C12.4.
Students think that the mass of the particles in a substance changes as the temperature changes.	To address this, place ice in a dish on a balance. Record the mass as the ice melts and the water starts to warm. See key idea card C12.5.
Students think that the molecules of a substance break down into individual atoms when a substance evaporates. During evaporation, water breaks down into hydrogen and oxygen.	To address this, use molecular models while discussing change of states (or boil some water in front of the students). See key idea card C12.5.
Students think that a kilogram of feathers is lighter that a kilogram of lead.	To address this, we need to understand density which is mass/volume. A kilogram of lead has a very small volume compared to that of feathers, so is more dense and just appears to be heavier. See key idea card C12.3.

CONNECT

The student should watch the video on the powerful idea for the session. The powerful idea for Session 12 is the particle model of matter.

A key idea is that the properties of a material are determined by how the particles are arranged.

Teaching tips

Exam questions on the particle model are often integrated as part of a question on a different topic. Students may be asked to:

- draw or interpret particle diagrams showing substances in the solid, liquid or gaseous state
- compare the properties of solids, liquids and gases
- determine the state of a substance using the data provided in the question
- describe how the arrangement of the particles in a substance changes during a change of state
- discuss reasons why different substances have different densities.

As all materials are made from particles, a wide variety of different contexts may be used to ask questions on the particle theory. Encourage students to learn the basic principle that the type and strength of bonds and forces between the particles determine the properties of the material. Once this piece of knowledge is secure, students just need to apply it to the different contexts.

SOLVE

Now turn back to the original question with the student. Work through the writing frame in the workbook together. The solutions are given below.

The table shows the density of some different materials.

Substance	Density in kg/m³
Iron	8000
Gold	19000
Water	1000
Air	1.3

a) Explain why large molecules generally have higher melting points than smaller molecules.

Use what you know about energy and changing state in your answer.

b) Explain why different materials have different densities.

Use data from the table and your knowledge of the particle theory to support your answer.

Draw the particles in solids, liquids and gases.

See key ideas card C12.2.

What is holding the particles in the patterns you have drawn?

Forces of attraction.

For part a): The particles you have drawn represent molecules. With bigger molecules, do they have more or less force holding them together?

They have more force holding them together.

For part a): When a substance is heated, the energy makes the substance warmer. When the substance changes state, the energy is used to break the forces that hold the particles of the substance together. What happens to the temperature when a substance changes state?

The temperature stays the same or remains constant.

For part a): Now bring together what you have written to explain why bigger molecules melt at higher temperatures.

Bigger molecules melt at higher temperature because it takes more internal energy to break the forces of attraction between the molecules. We can increase the thermal energy of the particles by heating.

Therefore, bigger molecules have a higher melting point.

For part b): Look at the data in the table. Write the state of each substance at room temperature (solid, liquid or gas).

Iron: solid. Gold: solid. Water: liquid. Air: gas.

For part b): Write down the formula that links density with mass and volume. How does particle size affect the density of a substance?

$$\text{density} = \frac{\text{mass}}{\text{volume}}$$

The larger the particle, the greater the density.

For part b): Now bring together what you have written to explain why different materials have different densities.

The density of a material depends on the size of the particles and how closely they are packed. So most liquids are less dense than solids and more dense than gases. The particles in a solid are packed closer together than in a liquid or gas.

PRACTISE

Use the additional practice questions on this powerful idea at the back of the workbook in future sessions. Additional practice questions for Session 12 can be found in the Determine, Evaluate and Predict sections.

SESSION 13: ATOMIC STRUCTURE

THINK

Ask the student to read the question for Session 13 in the workbook and scribble down what they know on that topic. Now ask the student to have a go at the questions on Adapt© for Session 13.

Common misconceptions

Students often think that electrons and protons are the only fundamental particles in an atom.	To address this, when drawing dot and cross diagrams, encourage students to include the number of protons and neutrons in the nucleus, before adding the electrons into the electron shells. See key idea card C13.2.
Students can find it difficult to distinguish the nucleus of an atom from the nucleus of a cell.	To address this, use specific language and explain the context in which you are using the word, so students link meaning to context. See key idea card C13.2
Students do not always understand the significance of the subscript in a chemical formula.	To address this, present students with models of chemical compounds and ask them to count the atoms before writing down the formula. Repeat the activity, but starting with the chemical formula, and ask students to identify the number and types of atoms present. See key idea cards C13.6 and C13.7.
Students often confuse the terms atomic number and relative atomic mass. They think that in the periodic table the elements are arranged in order of increasing atomic mass.	To address this, clarify the definitions and ask students to find all instances on the periodic table where elements are not arranged by increasing atomic mass. See key idea cards C13.4 and C13.5.
Students often think that in an atom the space between the nucleus and the electron shells is full of air. They also find it difficult to visualise the relative size of electrons to protons and neutrons.	To address this, explain that air is the mixture of gases found in the atmosphere. Air cannot get into an atom. The space between the nucleus and the electron shells is empty.

CONNECT

The student should watch the video on the powerful idea for the session. The powerful idea for Session 13 is structure of the atom.

In an atom, protons and neutrons are found in the nucleus which is surrounded by electrons in different electron shells. All elements are listed in the periodic table which is arranged so that you can see patterns in the types and strength of the reactions of each element relative to each other.

Teaching tips

Exam questions on atomic structure typically ask students to:

- identify the number of protons, neutrons and electrons given the atomic number and the mass number
- draw or label a diagram of a specific atom
- state the relative mass and charges of the subatomic particles
- compare the physical and/or chemical properties of a group of elements and relate the trends in properties to the atomic structure
- describe how the model of the atom has changed over time.

Frequently, several of these ideas are combined together in one question.

It is a good idea to remind students that the key information needed to answer the question is usually given at the start of the question. Encourage students to highlight the relevant pieces of information as they read the question. Sometimes, they may need to extract further information from the periodic table. In these questions, students need to be able to apply their learning about the structure of the atoms and how atoms are arranged in the periodic table. So, you can help students to feel more confident about this topic by making sure they are familiar with the layout of the periodic table.

SOLVE

Now turn back to the original question with the student. Work through the writing frame in the workbook together. The solutions are given below.

The trends in physical and chemical properties in the periodic table depend on atomic structure and electronic structure.

Compare the reactivity of magnesium and calcium.

Find magnesium and calcium in the periodic table. Which group are they in? What does this mean?

They are both in Group 2. This means they both have two electrons in their outer shell.

How many protons, neutrons and electrons does an atom of magnesium have? How many protons, neutrons and electrons does an atom of calcium have? Use the periodic table to find out. Write your answers in the table.

	No. of protons	No. of neutrons	No. of electrons
Magnesium	12	12	12
Calcium	20	20	20

Draw the electronic structure of calcium and magnesium. Use crosses for electrons. Label the energy levels.

Magnesium:

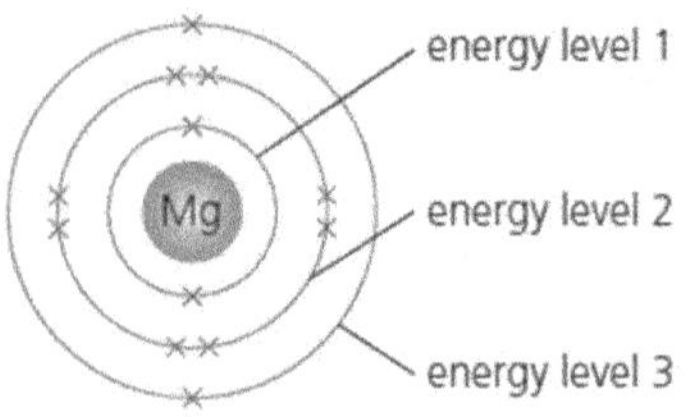

Calcium:

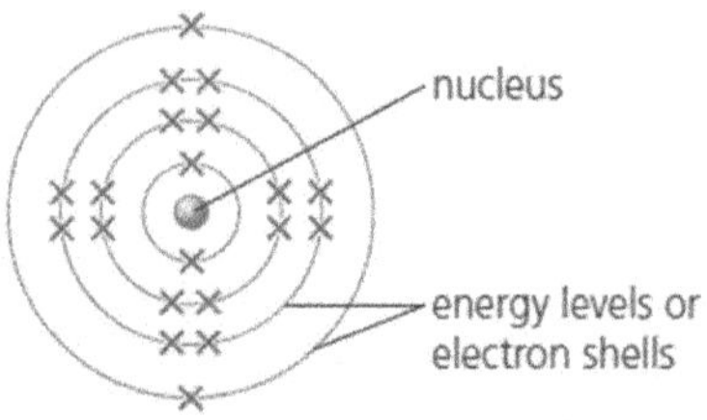

Magnesium is above calcium in the periodic table. Which one is more reactive? Why?

Calcium is more reactive. This is because, during chemical reactions, the outer electrons are 'transferred'. Down the group, the outer electrons are further away from the nucleus. Therefore, the electrostatic attraction with the nucleus is weaker and so it is easier to transfer electrons during a chemical reaction.

Bring together your previous answers to make a list of how magnesium and calcium are similar.

- Both are metals.
- Both are in Group 2 of the periodic table.
- Both have 2 electrons in their outer shell.

Bring together your previous answers to make a list of how magnesium and calcium are different.

- Calcium has more electron shells or energy levels.
- In calcium, the outer shell electrons are further away from the nucleus, so the electrostatic attraction with the nucleus is weaker.
- Calcium is therefore more reactive than magnesium.

PRACTISE

Use the additional practice questions on this powerful idea at the back of the workbook in future sessions. Additional practice questions for Session 13 can be found in the Describe and Determine sections.

SESSION 14: CALCULATING CHEMICAL CHANGE

THINK

Ask the student to read the question for Session 14 in the workbook and scribble down what they know on that topic. Now ask the student to have a go at the questions on Adapt© for Session 14.

Common misconceptions

Students think that the atoms of the reactants of a chemical reaction are transformed into other atoms.	To address this, use models or diagrams to represent the atoms in the reactants. Show that the products contain the same atoms, just in a different arrangement. See key idea cards C14.2 and C14.4.
Students think that mass is lost during a chemical reaction because the weight decreases.	To address this, reinforce the idea that a gas does have mass. Show that the mass will stay the same if you capture that gas in a balloon. See key idea card C14.5.
Students think that the total mass increases in a precipitation reaction because the precipitate produced is solid and it is heavier than a liquid.	To address this, reinforce the idea that during a chemical reaction the atoms in the reactants are rearranged. Nothing is added in. Carry out a precipitation reaction on a balance. See key idea card C14.4.
Students think that chemical equations are balanced by changing the subscripts in the chemical formula.	To address this, draw a box around each formula in the equation. When balancing equations you can only change the number of boxes and not what is written inside them. See key idea card C14.3.

CONNECT

The student should watch the video on the powerful idea for the session. The powerful idea for Session 14 is that during a chemical reaction, atoms are rearranged and new products are formed.

A key idea is that we can calculate the number of reactants and products because no atoms are lost during the reaction.

Teaching tips

Exam questions on calculating chemical change typically ask students to:

- complete a word equation or balance a symbol equation
- describe or determine what type of chemical reaction has taken place
- apply the law of conservation of mass to a given reaction
- use the data provided to calculate either the mass of a reactant or product or the relative formula mass.

Frequently, several of these ideas are combined together in one question. Students may also be asked to solve problems that require them to either scale up or scale down the quantities used in a chemical reaction.

Students will normally find the key information required to answer the question given at the start of the question, or they will be asked to look at the periodic table. Sometimes they will be asked to extract and use information from either a table, pie or bar chart, or line graph.

They need to be familiar with the chemical symbols but do not need to memorise the atomic or mass numbers, as a periodic table is always provided.

The examples used to calculate chemical change may be taken from anywhere in the specification. This could include any type of reaction, from acids and alkalis to thermal decomposition to extractions of metals.

SOLVE

Now turn back to the original question with the student. Work through the writing frame in the workbook together. The solutions are given below.

Magnesium (a metal) reacts with oxygen (a gas) to form a compound (a white powder).

a) Predict whether the solid product would be heavier, lighter or the same mass as the solid reactants.

Use an equation to explain your prediction.

b) Calculate how much magnesium oxide (MgO) would be made by the combustion of 5 g of magnesium (Mg).

For part a): Decide what the reactants and products of the reaction are. Write a word equation.

Reactants: magnesium and oxygen. Products: magnesium oxide.

magnesium + oxygen → magnesium oxide

For part a): Now write a formula equation.

The unbalanced equation is: $Mg + O_2 \rightarrow MgO$

For part a): Is the formula equation balanced? In other words, are there the same number of atoms of each element on both sides of the equation? If not, balance it.

$2Mg + O_2 \rightarrow 2MgO$

For part a): Now look at the equation again. Which reactants and which products are solids?

Reactant: Mg. Product: MgO.

Now look at the balanced equation to compare how many atoms of the solid Mg (2) react to form how many molecules of the solid MgO (2). You can now predict whether the mass of product (2MgO) is heavier, lighter or the same mass as the magnesium reacted (2Mg).

The oxygen reactant is a gas, so the mass of reactant we need to compare is that of the solid magnesium only. The equation tells us that there is only one product, which is solid. Therefore, the mass of solid product (MgO) will be heavier than the mass of solid reactant (Mg).

For part b): The relative atomic mass of Mg is 24 and O is 16. Calculate the relative formula mass of MgO.

24 + 16 = 40

For part b): Using the equation $2Mg + O_2 \rightarrow 2MgO$, if we start with 24 g of Mg, how much MgO will be made?

40 g

For part b): When we start with 5 g of magnesium, the mass of magnesium oxide produced is scaled down by the same proportion. Work out the multiplier.

$\frac{5}{24} = 0.21$

For part b): Now bring together your previous answers to work out how much MgO is formed from 5 g of Mg.

40 × 0.21 = 8.4 g

PRACTISE

Use the additional practice questions on this powerful idea at the back of the workbook in future sessions. Additional practice questions for Session 14 can be found in the Balance, Calculate, Define and Justify sections.

SESSION 15: BONDING

THINK

Ask the student to read the question for Session 15 in the workbook and scribble down what they know on that topic. Now ask the student to have a go at the questions on Adapt© for Session 15.

Common misconceptions

Students think that atoms want to form bonds or need to obtain a full outer shell.	To address this, encourage students to think about the elements and their position in the periodic table. Avoid anthropomorphic analogies such as 'finding a partner' to bond to. Atoms don't make decisions! See key idea card C15.1.
Students think that the ionic bond is the transfer of electrons from one atom to another.	To address this, explain that the ionic bond is the electrostatic attraction between a positive and a negative ion. Demonstrate electrostatic attraction by rubbing a balloon on a jumper and then 'sticking it' to the wall. The transfer of electrons is the process of forming the ions and thus of the formation of the ionic bond. See key idea cards C15.2 and C15.3.
Students think that covalent molecules have weak bonds.	To address this, address the difference between strong covalent bonds between the atoms in the molecule and weak intermolecular forces between the molecules. The latter explains why covalent molecules have low melting points. Use models to illustrate this. See key idea card C15.7.
Students think that there is no bonding in metals.	To address this, discuss the properties of metals. Metals are good conductors of electricity. So, there must be electrons free to move. If there was no bonding, there would be no free electrons as they would all be attached to individual atoms. See key idea card C15.6.

CONNECT

The student should watch the video on the powerful idea for the session. The powerful idea for Session 15 is chemical bonds are formed when atoms react together by either sharing, donating or accepting electrons.

A key idea is that how particles are arranged depends on the strength of the bonds and forces that hold the particles in position.

Teaching tips

Generally, exam questions on chemical bonding fall into two main categories. Questions that ask students to:

- identify the type of bonding present in a range of different materials, or compare and explain the physical and chemical properties of different materials
- describe and explain how ionic or covalent bonds between two or more atoms are formed; this often includes either drawing or completing dot and cross diagrams.

At the start of exam questions, students are often provided with a 2D or 3D image of the models used to describe the bonding in a substance. They must be able to recognise these models as a metallic or crystal lattice, or as simple or giant covalent molecules.

Questions about structure and bonding may appear as part of a question on another topic, such as electrolysis or states of matter. Students may be asked to look at a table of data such as melting points or density and to use the information to suggest the type of bonding present in a substance.

SOLVE

Now turn back to the original question with the student. Work through the writing frame in the workbook together. The solutions are given below.

Element	Atomic number	Relative atomic mass	State at room temperature
Oxygen	8	16	Gas
Sodium	11	23	Solid
Chlorine	17	35.5	Gas
Sodium oxide			Solid

The table gives some information about the atoms of sodium, oxygen and chlorine.

a) Suggest how sodium oxide crystals and molecules of chlorine gas are formed.

b) Explain why sodium oxide is a solid but chlorine is a gas.

Use what you know about ionic bonding and covalent bonding in your answer.

For part a): Ionic bonds form when electrons are lost by one atom and given to another. What happens to the charge of an atom when an electron is lost? What happens to the charge of an atom when an electron is gained?

When an atom loses an electron, it is left with a positive (+) charge.

When an atom gains an electron, it gains a negative (−) charge.

For part a): How many electrons has each sodium atom lost?

One.

For part a): How many electrons has each oxygen atom gained?

Two.

For part a): How many sodium ions are needed for each oxygen atom? Write this number next to the sodium in the formula for sodium oxide. Make sure the N of sodium is capitalised and the a is lower case. The O of oxygen is a capital.

Na_2O

For part a): Now write a sentence about what happens when an ionic bond is formed between sodium and oxygen.

During the formation of the ionic bond, each oxygen atom gains two electrons to form a 2− ion and each sodium atom loses an electron to form a + ion. The ionic bond is the electrostatic attraction between the ions.

For part a): Now bring together your ideas to write a sentence about what happens when chlorine atoms form covalent bonds to make a chlorine molecule.

The two chlorine atoms each share 1 electron from their outer shell.

For part b): Now bring together what you have written and try to explain why sodium oxide is a solid but chlorine is a gas.

Sodium oxide is a solid as the ions form a giant crystal lattice. It takes a lot of energy to break the forces of attraction between all the ions. Chlorine forms simple molecules made from two chlorine atoms. The forces between the molecules are very small.

PRACTISE

Use the additional practice questions on this powerful idea at the back of the workbook in future sessions. Additional practice questions for Session 15 can be found in the Draw section.

SESSION 16: PROPERTIES OF MATERIALS

THINK

Ask the student to read the question for Session 16 in the workbook and scribble down what they know on that topic. Now ask the student to have a go at the questions on Adapt© for Session 16.

Common misconceptions

Students think that ionic compounds form discrete molecules.	To address this, use 3D models to show the difference between a CO_2 molecule and an NaCl crystal lattice. See key idea card C16.3.
Students think that an alloy is a compound of two different metals.	To address this, explain that all compounds have a chemical formula, such as CO_2 or C_2H_5OH, but alloys do not have a chemical formula. This is because they are a mixture. Brass is a mixture, or alloy, of copper and zinc. See key idea card C16.5.
Students think that all metals are magnetic.	To address this, hold a magnet near an aluminium drink can and a steel can. The aluminium will not be attracted to the can but the steel one will. Only iron, nickel and cobalt are magnetic. Note that steel is an alloy of iron.
Students think that graphite cannot conduct electricity because it is a non-metal that forms a giant covalent molecule.	To address this, first acknowledge that graphite is the exception to the rule! Show students that it does conduct electricity by clipping a pencil into an electric circuit. You will need to sharpen both ends. Then, deduce that there must be some electrons free to move, before looking at the actual structure. See key idea card C16.7.

CONNECT

The student should watch the video on the powerful idea for the session. The powerful idea for Session 16 is the properties of a material depend on how the atoms are bonded, the size of the molecules and the arrangement of the atoms and ions.

A key idea is that materials will only conduct electricity if electrons are free to move.

Teaching tips

Exam questions on properties of materials often involve some data analysis followed by an explanation as to why a material has a particular property. These questions are often linked to questions on different types of bonding, so there is some overlap with Session 15. Students may be asked to look at a table of data or a graph, and to use the information to suggest why a material has specific properties. Students will be expected to discuss properties such as melting and boiling points, electrical conductivity and solubility in terms of the weak forces between molecules, ionic lattices and delocalised electrons.

Questions about properties of materials may appear as part of a question on another topic, such as electrolysis, states of matter or chemical bonding. Students may be asked to apply their knowledge to explain the properties of materials they have not met before. Encourage students to carefully study all the information provided in the question before they attempt to answer it.

SOLVE

Now turn back to the original question with the student. Work through the writing frame in the workbook together. The solutions are given below.

Graphite has two helpful properties: It conducts electricity and it is used as a solid lubricant.

a) Explain why graphite has both properties.

b) Diamond is another form of carbon. Explain why diamond is used for drill tips whereas graphite is not.

Use what you know about how the carbon is bonded in each substance in your answer.

For part a): Carbon is in Group 4 of the periodic table. How many electrons does it have available for bonding in its outer shell?

Four electrons.

For part a): In graphite, each carbon atom only uses three of its available electrons for bonding, and has one free electron left over. The diagram below shows the structure of graphite. Add labels to the diagram to show the types of bond and the atoms present. What is the name given to the free electrons?

See key ideas card C16.7.

The free electrons are called delocalised electrons.

For part a): Which of the features you have labelled makes graphite a solid lubricant? Explain this.

The weak intermolecular forces between layers of carbon atoms in graphite make graphite a solid lubricant.

For part a): Which of the features you have labelled makes graphite a conductor of electricity? Explain this.

The delocalised electrons make graphite able to conduct electricity.

For part b): What physical property does a material need if it is going to be used as a drill tip?

It must be very hard.

For part b): Suggest a reason why graphite is soft.

The weak intermolecular forces between the graphite layers mean they are easy to separate.

For part b): Diamond forms a giant molecular structure. Why is diamond harder than graphite?

Diamond uses all four of its available electrons for bonding. Every carbon atom is joined to four other carbon atoms by strong covalent bonds. This makes the structure very hard.

PRACTISE

Use the additional practice questions on this powerful idea at the back of the workbook in future sessions. Additional practice questions for Session 16 can be found in the Evaluate and Justify sections.

SESSION 17: ELEMENTS, COMPOUNDS AND MIXTURES

THINK

Ask the student to read the question for Session 17 in the workbook and scribble down what they know on that topic. Now ask the student to have a go at the questions on Adapt© for Session 17.

Common misconceptions

Students think that a compound is formed when two or more of the same elements are chemically joined together.	To address this, use precise language and definitions and make sure students get plenty of practice. ('A compound is two or more different elements …') See key idea cards C17.1 and C17.2.
Students think that an element is a substance made from one atom.	To address this, use precise language and definitions and make sure students get plenty of practice. ('An element is one type of atom.') See key idea cards C17.1 and C17.2.
Students think that filtration can separate solutes from solvents.	To address this, pour a salt solution through a filter and show that nothing has been left on the filter paper. Repeat with some muddy water and see the separation. Clearly define the terms soluble, insoluble, solute, solvent and solution. See key idea card C17.7.
Students think that incorrectly measuring the distance a spot has travelled during chromatography.	To address this, work through an example with the class. Use a ruler to find the centre of the spot and also draw in the solvent front. See key idea card C17.5.
Students think that during chromatography, the colours come from a chemical reaction with the water.	To address this, stress that separation techniques use physical properties to separate a mixture. No chemical change takes place. In this case the mixture dissolves in the solvent. The different components then travel up the chromatography paper. See key idea card C17.5.

CONNECT

The student should watch the video on the powerful idea for the session. The powerful idea for Session 17 is that elements are substances made out of only one type of atom. Mixtures are made out of two or more substances that are not chemically bonded together. Compounds are substances that contain at least two different elements chemically bonded together.

A key idea is that the parts of a mixture can be separated using the different properties of each part of the mixture.

Teaching tips

Exam questions on elements, mixtures and compounds typically ask students to:

- use particle diagrams to identify whether a substance is an element, compound or mixture
- compare the different properties of two elements, or compounds
- write chemical equations as a way of showing how elements combine to become a compound during a chemical reaction
- describe how different physical properties are used to separate a mixture into its components using filtration, crystallisation, distillation or chromatography.

Several of these ideas may be combined into a practical-based question, as separation techniques are used in three of the required practicals: making salts, chromatography and water purification. In questions about chromatography, students may have to interpret chromatograms and calculate R_f values. A range of different contexts may be used to set the scene.

SOLVE

Now turn back to the original question with the student. Work through the writing frame in the workbook together. The solutions are given below.

Chromatography is a technique used to separate mixtures of different compounds or to test a compound for purity.

The diagram shows a chromatogram.

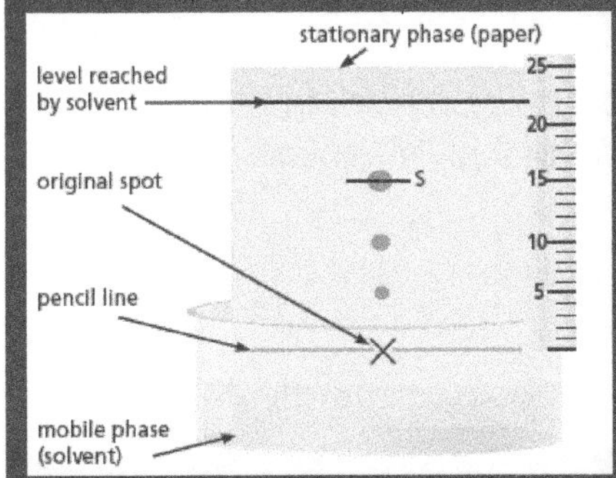

a) Describe how chromatography is used to identify the dyes used in food colouring.

b) Calculate the R_f value of the spot labelled S on the diagram.

For part a): Chromatography is used to separate mixtures. What is a mixture?

A mixture is a substance containing two or more different elements or compounds that are not chemically bonded together. They can be separated by physical processes.

For part a): Which physical property is used in chromatography to do the separation?

Solubility.

For part a): What is meant by the terms mobile phase and stationary phase?

The mobile phase is the solvent that the mixture dissolves in. The stationary phase is the chromatography paper that the solvent travels up.

For part a): Why is a pencil used to draw the starting line?

Pencil lead is insoluble (it does not dissolve in water).

For part a): Look at the diagram showing the chromatogram. How many dyes were found in the sample?

Three dyes (as there are three spots).

For part a): We can identify the dyes by calculating the R_f *value. Write down the equation that links* R_f *values to the distance travelled by the spot.*

$$R_f = \frac{\text{distance travelled by the spot}}{\text{distance travelled by the solvent}}$$

For part a): Now bring together these ideas to describe how chromatography is used to identify the dyes used in food colouring.

- Draw a pencil line on some chromatography paper.
- Spot the sample onto the line.
- Hang the paper in the chromatography tank.
- Let the solvent move up the paper.
- Leave the paper to dry.
- Calculate the R_f value.
- Compare the values to those of known dyes.

For part b): Calculate the R_f *value of the spot labelled S on the diagram.*

$$R_f = \frac{15}{22} = 0.68$$

PRACTISE

Use the additional practice questions on this powerful idea at the back of the workbook in future sessions. An additional practice question for Session 17 can be found in the Justify section.

SESSION 18: EARTH CHEMISTRY

THINK

Ask the student to read the question for Session 18 in the workbook and scribble down what they know on that topic. Now ask the student to have a go at the questions on Adapt© for Session 18.

Common misconceptions

Students think that crude oil is a compound.	To address this, explain that it is a mixture of hydrocarbon compounds. Refer students to the fractional distillation tower image, and remind them that fractional distillation is a separation technique, used to separate a mixture of liquids. You could go on to look at what the different fractions are used for. See key idea cards C18.1 and 18.7.
Students think that the process of cracking is often confused with fractional distillation.	To address this, stress the idea that when something is cracked it is broken into smaller pieces. During cracking we break large hydrocarbon molecules into smaller ones. In fractional distillation, we are separating a mixture. See key idea cards C18.5 and C18.7.
Students think that the products of combustion are carbon dioxide, water and energy.	To address this, look at the chemical equation. There are no atoms for energy, therefore it cannot be a product. In this reaction energy is released in the form of heat, light and sound. See key idea card C18.2.
Students think that general confusion between the processes of photosynthesis, respiration and breathing when being applied to the formation of the Earth's atmosphere.	To address this, cross-reference to what students have learned in biology. Ensure that students can recall the meaning of each word, then provide practice with using these words correctly in different contexts. See key idea cards C18.3 and C18.4.

CONNECT

The student should watch the video on the powerful idea for the session. The powerful idea for Session 18 is that the Earth's atmosphere was formed millions of years ago from chemical reactions and biological processes. Crude oil is a fossil fuel made from the remains of organisms that lived millions of years ago.

A key idea is that crude oils are a mixture of hydrocarbon molecules which can be used to make many different useful materials including polymers, plastics and detergents.

Teaching tips

Exam questions on Earth chemistry include questions about:

- the formation of crude oil and its separation into different groups of hydrocarbons by fractional distillation
- how fractional distillation works, linking the size of the molecules to different fractions
- the complete and incomplete combustion of hydrocarbon fuels and the effect these have on the atmosphere
- how the Earth's atmosphere has evolved over time
- how many useful materials, including polymers and plastics are formed, from crude oil.

Students may be asked to:

- label diagrams, e.g. a fractional distillation tower
- use models to describe how polymers are formed, or name individual hydrocarbon molecules
- interpret data relating to the composition of the atmosphere from tables and charts
- balance chemical equations.

Questions about the atmosphere may also refer to the carbon cycle and to processes students learn about in biology, including respiration and photosynthesis. A variety of different contexts may be used when referring to useful products made from crude oil.

SOLVE

Now turn back to the original question with the student. Work through the writing frame in the workbook together. The solutions are given below.

Earth's atmosphere was formed from chemical reactions and biological processes taking place over billions of years.

Crude oil is a fossil fuel formed millions of years ago.

Compare how photosynthesis and burning fossil fuels have changed the Earth's atmosphere.

What is photosynthesis? Describe photosynthesis in words or write the equation.

Plants make glucose from carbon dioxide and water by photosynthesis and produce oxygen gas as a by-product.

carbon dioxide + water → glucose + oxygen

What is crude oil made from and where was it made?

Crude oil is made of the remains of dead organisms/sea animals. Crude oil was formed under the oceans.

Crude oil is a mixture of different substances. Describe how fractional distillation is used to separate them.

Crude oil is heated in a fractionating column until it forms a vapour. The vapour rises up the tower. The different hydrocarbons condense at the place in the column just below their boiling point. Short-chain hydrocarbons have lower boiling points than long-chain hydrocarbons, and so rise further up the tower. The different fractions are then collected.

What is combustion? Describe combustion in words or write the equation.

Combustion is when a fuel burns in oxygen to form carbon dioxide and water, giving out lots of energy.

fuel + oxygen → carbon dioxide + water

Now think about the development of the Earth's atmosphere. When did photosynthesis start happening? When did combustion start happening?

Photosynthesis started when plants such as algae started to form billions of years ago, when the atmosphere had lots of carbon dioxide. Combustion started to happen when humans began to burn fossil fuels a few hundreds of years ago.

Now bring together your answers to compare the two processes. You might like to think about what each process added or removed from the atmosphere, and when each one happened.

Photosynthesis takes carbon dioxide out of the atmosphere and puts oxygen in, whereas combustion uses oxygen and produces carbon dioxide. The early atmosphere had lots of carbon dioxide which gradually got replaced with oxygen as more plants started to grow. In more recent times, the levels of carbon dioxide have started to increase as more fossil fuels have been burned.

PRACTISE

Use the additional practice questions on this powerful idea at the back of the workbook in future sessions. An additional practice question for Session 18 can be found in the Suggest section.

SESSION 19: ELECTROLYSIS

THINK

Ask the student to read the question for Session 19 in the workbook and scribble down what they know on that topic. Now ask the student to have a go at the questions on Adapt© for Session 19.

Common misconceptions

Students think that a metal is always produced during electrolysis.	To address this, remind students that this is only true for molten compounds and not aqueous solutions. Provide students with a set of rules covering what is produced at each electrode during the electrolysis of an aqueous salt solution. Ask questions that will allow them to apply the rules. See key idea card C19.4.
Students think that aluminium is produced when aluminium oxide reacts with cryolite.	To address this, discuss why cryolite is added – to lower the melting point of aluminium oxide. To support understanding, use the example of adding salt to roads in the winter to lower the melting point of water, to stop it from freezing. Impurities lower the melting point of pure substances. See key idea card C19.6.
Students often confuse the terms anode and cathode.	To address this, use a phrase or mnemonic to help remember that the anode is positively charged. For example, + means add. A is the first letter of anode. Encourage students to make up their own. That way they will remember it. See key idea card C19.1
Students think that electrolysis works because the electric current moves through the solution.	To address this, remind students that an electric current is a flow of electrons. Electrons are found in atoms. Metals conduct electricity because the delocalised electrons can move. In electrolysis it is the positive and negative ions that are moving through the solution towards the electrodes where the electrons are either taken from or given to the metals. See key idea cards C19.3 and C19.5.

CONNECT

The student should watch the video on the powerful idea for the session. The powerful idea for Session 19 is electrolysis.

A key idea is that opposites attract. During electrolysis the positive ions are attracted to the negative cathode and the negative ions are attracted to the positive anode.

Teaching tips

Exam questions on electrolysis include questions about the products formed during electrolysis on both a small scale in the laboratory and on a large industrial scale. Students may be asked about the electrolysis of molten compounds and aqueous solutions. A good answer will include an explanation of how the products are formed.

In order to work out what is formed at each electrode during the electrolysis of an aqueous solution, students will need to learn and apply the set of rules. Encourage students to start by noting down all the ions present in the solution; remind them to include the H^+ and OH^- ions from the water.

During the course, students will have learned about how aluminium is extracted from its ore by electrolysis. In an exam question, information may be supplied about the extraction of a different reactive metal such as zinc. Students will need to apply the general principles, such as negative ions are attracted to the positive electrode (anode) and positive ions are attracted to the negative electrode (cathode), to the new context. Give students plenty of practice of identifying the ions in the extraction of different metals to help with this.

SOLVE

Now turn back to the original question with the student. Work through the writing frame in the workbook together. The solutions are given below.

a) Describe how pure aluminium is extracted from aluminium ore.

b) Explain why aluminium is so expensive.

For part a), Where is aluminium found naturally?

Aluminium is found in the ground as aluminium ore or aluminium oxide.

For part a): Why can aluminium not be extracted using carbon like iron can?

Aluminium is too reactive.

For part a): What are the names of the two electrodes used in electrolysis?

Positive electrode: anode. Negative electrode: cathode.

For part a): Add labels to the diagram to show what happens at each electrode in electrolysis.

See key ideas card C19.5.

For part a): Describe or draw how electrolysis is used to extract aluminium. Add as many labels as you can.

See key ideas card C19.6.

- Aluminium oxide is mixed with cryolite and heated until it melts.
- Cryolite lowers the melting point.
- An electric current is then passed through the liquid.
- Al^{3+} ions are attracted towards the cathode, where the molten aluminium forms, which is then tapped off.
- The oxide ions are attracted to the anode, where oxygen gas is formed.

For part b): Give two reasons why a lot of energy is used during the extraction of aluminium.

- To melt the aluminium oxide.
- Electricity is needed for electrolysis itself.

For part b): Why do the anodes need to be replaced regularly?

As soon as the oxygen is formed it reacts with the carbon to form carbon dioxide, removing the carbon from the anode.

For part b): Now bring it together: Why is aluminium so expensive?

It takes a lot of energy to produce aluminium. First, to melt the aluminium oxide, and then for the electrolysis itself. The carbon anodes need to be replaced regularly because as soon as the oxygen is formed it reacts with the carbon to form carbon dioxide; so the carbon anodes need to be replaced, which adds to the expense.

PRACTISE

Use the additional practice questions on this powerful idea at the back of the workbook in future sessions. An additional practice question for Session 19 can be found in the Evaluate section.

SESSION 20: ACIDS AND ALKALIS

THINK

Ask the student to read the question for Session 20 in the workbook and scribble down what they know on that topic. Now ask the student to have a go at the questions on Adapt© for Session 20.

Common misconceptions

Students think that, when using indicators, acids are red and alkalis are blue.	To address this, show students a range of different indictors. Not all indicators show red for acid and blue for alkali. The pH of an acid is always less than 7, and the pH of an alkali is always greater than 7. See key idea card C20.2.
Students name salts incorrectly.	To address this, provide students with plenty of practice in naming salts by completing word equations. Then go on to ask them to identify the reactants they would need to produce salt X or Y, before writing more word equations. See key idea card C20.4.
Students write the formulae for acids and alkalis incorrectly.	To address this, build formulae into retrieval practice. Students need a working knowledge of this. See key idea card C20.5.
Students struggle to know where to start when asked how to make a sample of a soluble salt other than copper sulfate.	To address this, present the practical as a sequence of different processes that is used to make all soluble salts. Ask students to identify the reactants they need to make the salt, and then slot the names of reactants into the sequence. See key idea cards C20.4 and C20.6.
Students think that an acid is broken down during neutralisation.	To address this, take care to use correct scientific language. Provide students with clear definitions of acids as containing H^+ ions and alkalis containing OH^- ions. Most students will know that pure water is neutral and has a pH of 7. Use this fact to stress that neutralisation is a chemical reaction and occurs when: $H^+ + OH^- \rightarrow H_2O$ See key idea card C20.5.

CONNECT

The student should watch the video on the powerful idea for the session. The powerful idea for Session 20 is the reactions of acids and alkalis.

A key idea is that when an acid reacts with an alkali, the hydrogen and hydroxide ions rearrange to make water.

Teaching tips

Exam questions on acids and alkalis will often ask students to write or complete word and/or symbol equations for neutralisation reactions. Encourage students to learn the general equations and then apply them to the reactants given in the questions. Two of the required practical activities – making a soluble salt and temperature changes – require students to use and apply their knowledge of acids and alkalis. Therefore, it is also common to get practical-based questions integrated into this topic, together with ideas about the conservation of mass or energy changes.

Exam questions about acids and alkalis often use a range of everyday contexts, such as:

- the effects of acid rain
- neutralising acidic soil
- indigestion and the use of antacids

how to avoid tooth decay by not drinking too many acidic drinks.

SOLVE

Now turn back to the original question with the student. Work through the writing frame in the workbook together. The solutions are given below.

Acids and alkalis are often used in homes in cooking, cleaning and medicines.

Sodium bicarbonate reacts with acids, neutralising them and producing a salt.

Explain why sodium bicarbonate is used in indigestion tablets.

Determine the reactants and products.

What is an acid?

A substance that produces H^+ ions in water; or, an aqueous solution of pH less than 7.

Why does extra stomach acid need treating?

Excess stomach acid may cause pain, such as heartburn.

The acid in the stomach can be neutralised by sodium bicarbonate. Write a word equation for the reaction.

acid + sodium bicarbonate → a salt + water + carbon dioxide

The formula for sodium bicarbonate is $NaHCO_3$. Hydrochloric acid is HCl. Write down the formula of the three products. Write how many H atoms there are in a molecule of water. How many C atoms are there in carbon dioxide?

Products: NaCl, CO_2, H_2O. There are 2 hydrogen atoms in water and 1 carbon atom in carbon dioxide.

Which product is a gas? What test would you carry out for the gas produced? Describe the test.

Carbon dioxide. Bubble the gas into limewater, and if it goes cloudy then carbon dioxide is present.

The salt produced is called sodium chloride. To practise naming other salts, what is the name of the salt made when hydrochloric acid reacts with magnesium carbonate? Write a word equation for the reaction.

Magnesium chloride.

hydrochloric acid + magnesium carbonate → magnesium chloride + water + carbon dioxide

Naming salt practice: What is the name of the salt produced when sulfuric acid reacts with sodium hydroxide? Write a word equation for the reaction.

Sodium sulfate.

sulfuric acid + sodium hydroxide → sodium sulfate + water

Now, bring these ideas together to explain why sodium carbonate is used in indigestion tablets and give the reactants and products.

The acid in the stomach can be neutralised by sodium bicarbonate.

acid + sodium bicarbonate → a salt + water + carbon dioxide

Reactants: stomach acid, sodium bicarbonate.

Products: a salt, water, carbon dioxide.

PRACTISE

Use the additional practice questions on this powerful idea at the back of the workbook in future sessions. Additional practice questions for Session 20 can be found in the Balance, Evaluate and Explain sections.

SESSION 21: RATES OF REACTION

THINK

Ask the student to read the question for Session 21 in the workbook and scribble down what they know on that topic. Now ask the student to have a go at the questions on Adapt© for Session 21.

Common misconceptions

Students think that when two particles collide a chemical reaction always takes place.	To address this, remind students of collision theory. The particles must have enough energy and must collide in the right orientation for a reaction to take place. If you hold a match in the air, oxygen particles will be bumping into it but, because they do not have sufficient energy, no reaction takes place, and the match does not burn. See key idea card C21.2.
Students think that the rate of a reaction can be increased by increasing the temperature, as this increases the frequency of collisions above the activation energy.	To address this, revisit the terms activation energy, frequency and collision. Once clearly defined, go on to discuss how, by increasing the temperature, we are increasing the energy of the particles, so that the activation energy is reached. See key idea cards C21.5 and C21.8.
Students think that the rate of reaction decreases when a powdered form of the substance is used.	To address this, demonstrate the concept of surface area by crushing a cube of sugar and spreading out the granules over the table. Explain that there are now more places when reactant particles can collide with the sugar particles because they cover more space. This will lead to an increase in successful collisions and rate of reaction. See key idea cards C21.3, C21.4 and 21.9.
Students think that at equilibrium the amounts of reactants and products are always the same.	To address this, explain that this is only true in some cases. However, the rate of reaction of the reactants and products is always the same at equilibrium. See key idea card C21.6.

CONNECT

The student should watch the video on the powerful idea for the session. The powerful idea for Session 21 is the rate of a chemical reaction and collision theory.

A key idea is that chemical reaction happens when particles collide with enough energy.

Teaching tips

Exam questions on rates of reaction will often be practical-based, focused around the processing and interpretation of experimental data and the application of collision theory. Students may be asked to:

- identify the different variables in an investigation
- draw a graph using the data provided in a table
- describe or compare the shape of graphs provided in the question
- complete a reaction profile of a chemical reaction with or without a catalyst being present
- use the collision theory to explain what happens to the particles during a chemical reaction
- explain how different factors including concentration, temperature and surface area affect the rate of reaction.

The practical example used in the question may be different from the experiment students have completed in class. Encourage students to learn the collision theory and provide opportunities for them to apply it to a range of different reactions.

SOLVE

Now turn back to the original question with the student. Work through the writing frame in the workbook together. The solutions are given below.

Hydrochloric acid reacts with magnesium to produce magnesium chloride and hydrogen gas.

The graph shows how the volume of hydrogen collected varies with time at two different temperatures of hydrochloric acid.

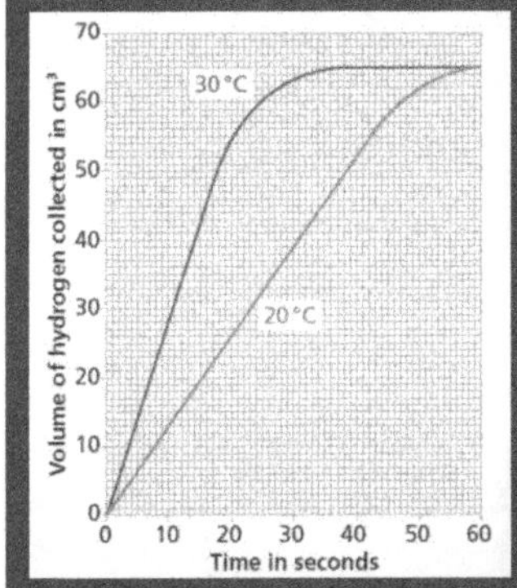

a) Use collision theory to explain why the reaction is faster when the acid is more concentrated.

b) Suggest another way of increasing the rate of reaction.

For part a): What happens to particles in a reaction?

The reactant particles must have a successful collision. This means they collide in the right direction and have enough energy to start the reaction.

For part a): What is the rate of a reaction?

It is the rate at which reactants are used up, or the rate at which products are made. It depends on the number of successful collisions that occur between molecules of the reactants per second.

For part a): What happens to the particles in a liquid when you heat it?

The kinetic energy of the particles increases.

For part a): What does this do to the rate of a reaction? Use the graph to support your answer.

The time taken for X amount of hydrogen to be produced decreases as the temperature is increased from 20 to 30 °C. For example, to produce 30 cm^3 of hydrogen gas: at 20 °C it takes 24 seconds and at 30 °C it takes 12 seconds. So the rate of reaction increases.

For part a): Bring all this together to explain why the rate of a reaction increases when the reactants are heated.

As the reactants are heated, the kinetic energy of the particles increases and they move more quickly. This means that they will collide with other particles more often, which will lead to in an increase in the number of successful collisions occurring each second, and so the rate of reaction increases.

For part b): Now list other ways a reaction can be speeded up.

- Increasing the concentration or pressure
- Increasing the surface area of the reactants.
- Using a catalyst.

PRACTISE

Use the additional practice questions on this powerful idea at the back of the workbook in future sessions. Additional practice questions for Session 21 can be found in the Justify and Suggest sections.

SESSION 22: CHEMISTRY PRACTICALS

THINK

Ask the student to read the question for Session 22 in the workbook and scribble down what they know on that topic. Now ask the student to have a go at the questions on Adapt© for Session 22.

Common misconceptions

Students are often confused about the difference between hazard and risk.	To address this, present students with clear definitions using examples from everyday life, e.g. hazards and risks involved when crossing a road. Then move on to chemistry-related examples. See key idea card C22.1.
Students are often confused about the difference between the independent and the dependent variable.	To address this, focus on literacy. What do the words dependent and independent mean? Once the meanings are established, apply the words to the 'working scientifically' context. See key idea card C22.2.
Students have problems choosing the correct equipment for measuring out a volume or mass accurately.	To address this, show how accuracy depends on the precision of the equipment. For example, weight of 2 g of sodium chloride on a balance is accurate only to 1 g. Take the sample and re-weigh it using a 1 d.p. balance and then a 2 d.p. balance and compare the actual readings. The same exercise could be repeated by measuring out water in different-sized measuring cylinders and beakers. See key idea card C22.3.
When measuring temperature, students often take the thermometer out of the reaction vessel to read the scale.	To address this, carry out a quick demonstration where you measure the temperature correctly and then incorrectly. Compare the two readings and explain why the first is correct and more accurate.
Students think that scientific ideas are well-established facts and do not change.	To address this, revisit the development of the periodic table and/or the atom. Emphasise how ideas changed when new experimental evidence no longer fitted the prevailing theory.

CONNECT

Teaching tips

Exam questions about the required practicals are used as mechanisms to ask questions about the range of practical skills outlined in the KS3 specification under 'Working scientifically – Experimental skills and investigations' and 'Working scientifically – Analysis and evaluation. Questions about health, safety and risk may be asked within any practical-based question. Students may be asked to recall the meaning of a hazard symbol and suggest what they can do minimise the risk.

Questions involving the preparation of soluble salts often ask students to plan or design an experiment to produce a sample of a named salt. They may be asked to produce a list of apparatus or draw a sequence of diagrams to show the different steps in the process. You can help students to feel more confident about this practical by making sure that they can identify the reactants needed to produce named salts and write the word/symbol equations.

Questions about investigating what happens when aqueous solutions are electrolysed using inert electrodes often ask students to predict the products formed at each electrode. They may then be asked to describe how to collect and test any gases formed at the electrodes. You can help students to feel more confident about this practical by making sure that they recall the rules for what is produced at the electrodes.

Investigating temperature changes in reacting solutions offers opportunities to plan an investigation by making a hypothesis, identifying the variables and writing a method. Students may also be asked to identify ways to minimise heat losses, such as use of an insulated container with a lid. You can help students to feel more confident about this practical by making sure that they know what a hypothesis is and how to identify independent, dependent and control variables.

The investigating rates of reaction experiment is frequently used to ask questions about data presentation or interpretation. Students are often presented with a table of data and asked to use the data to draw a graph. Or they are presented with a graph and asked about what the graph is showing and how this relates to what is happening during the reaction at a particle level. You can help students to feel more confident about this practical by getting them to practice drawing and answering questions from graphs.

Questions involving the analysis and purification of water may ask about the physical processes involved in separation techniques such as filtration, distillation and crystallisation. Students could be asked to describe or explain how each process works and/or correctly identify what type of impurity it can be used to remove. You can help students to feel more confident about this practical by ensuring that they can correctly name each piece of apparatus and understand its function.

Questions about chromatography often start by asking students to describe how a chromatogram is produced or to identify the errors in a given method. Questions may then go on to ask students to interpret a chromatogram by either comparing the spots or calculating an R_f value. Sometimes, this topic is used to examine whether a student can rearrange an equation, and so they are given the R_f value and asked to calculate how far the solvent front has travelled. You can help students to feel more confident about this practical by asking them to interpret a range of different chromatograms.

SOLVE

Now turn back to the original question with the student. Work through the writing frame in the workbook together. The solutions are given below.

Acids react with some metals to give a salt and hydrogen. Thermal energy is given out during the reaction.

Design an investigation to show how changing the concentration of hydrochloric acid affects the temperature change seen as it reacts with zinc metal.

What equipment would you use for this procedure? You can show this using a diagram. List the range of concentrations you would use. How will you measure the temperature of the reaction? When will you measure this?

Equipment: thermometer, balance, polystyrene cup and lid, 250 cm^3 beaker, 50 cm^3 and 10 cm^3 measuring cylinders, hydrochloric acid (1 mol/dm^3) and distilled water.

Use the thermometer to measure the temperature of the acid at the start and the end of the reaction.

List the steps of the practical procedure.

- Cut a 5 cm length of magnesium metal and clean it with sand paper until shiny.
- Measure out 30 cm^3 of acid and put it in the polystyrene cup. Stand the cup inside the beaker.
- Measure the temperature of the acid and record it in a table.
- Add the magnesium to the cup; put on the lid and gently stir with the thermometer through the hole.
- Watch the temperature rise. When the temperature stops rising, record the temperature in the table.
- Repeat the experiment 5 more times, but each time dilute the acid using distilled water. For example, in the second test use 25 cm^3 of acid and 5 cm^3 of water, etc.
- Finally, repeat each experiment and then work out the mean temperature change for each concentration of acid.

What are the hazards? What precautions would you take to avoid harm?

Dilute hydrochloric acid is an irritant to eyes and skin – wear safety glasses to stop splashes in the eyes; wash hands with water if acid gets on skin.

Magnesium is flammable – make sure there are no naked flames around.

Polystyrene cups can be knocked over easily – put the cup in a beaker to make it more stable.

What would you expect to happen?

As the concentration of the acid increases so does the temperature change of the solution. The temperature also goes up more quickly.

Why would you expect this to happen?

The reaction is exothermic, which means energy is given out as a result of successful collisions between the reactant particles. At lower concentrations all the acid will be used up and the magnesium will be in excess. This means there will be fewer successful collisions taking place. At higher concentrations there will be more successful collisions taking place and more energy is given out. The temperature will go up more quickly at the start because there will be more successful collisions per second in a more concentrated acid.

PRACTISE

Use the additional practice questions on this powerful idea at the back of the workbook in future sessions. Additional practice questions for Session 22 can be found in the Design, Draw, Plan, Show and Sketch sections.

SESSION 23: FORCES

THINK

Ask the student to read the question for Session 23 in the workbook and scribble down what they know on that topic. Now ask the student to have a go at the questions on Adapt© for Session 23.

Common misconceptions

Some students think that when an object is pushed, the force can become part of the object.	To address this, show a simple demonstration of blowing on a table-tennis ball, and discuss what happens to the air when it hits the ball. See key idea card P23.1.
Some students think that when two or more forces act on an object, the overall force is equal to the sum of all the numerical values of the forces.	To address this, demonstrate the effect of pulling a 1 kg mass with two newtonmeters, in opposite, and co-linear directions, and discuss the resultant motion of the 1 kg mass. See key idea cards P23.2, P23.3, P23.4 and P23.5.
Many students think that mass and weight are alternative words for the same property.	To address this, explain that mass is a measure of the amount of 'stuff' in an object, and weight is the force of gravity acting on the object. See key idea card P23.5.
Many students think that forces only act if two objects are in contact with each other.	To address this, show a simple demonstration of the (non-contact) interaction of two bar magnets, one free to spin on a watch glass, and the other hand held. See key idea card P23.7.
Many students think that if there is no resultant force acting on an object it will not move.	To address this, show a simple demonstration of a cup-cake case falling through air: the resultant force is zero, but it moves at a constant speed. See key idea cards P23.2, P23.3 and P23.5.

CONNECT

The student should watch the video on the powerful idea for the session. The powerful idea for Session 23 is that a force is a push, pull or twist that causes an object to change speed, change direction or change shape. A force can cause a stationary object to start moving or a moving object to become stationary. When two or more forces act on an object, their effects on the object are added together. If the forces cancel each other out they are said to be balanced, and there is no change in the movement or shape of the object. If the forces are unbalanced, then the movement or shape of the object will change. Newton's First and Second Laws of Motion explain how forces affect a single object. Newton's Third Law of Motion explains that when two objects interact the forces they exert on each other are equal and opposite.

A knowledge of the effect of forces is key to understanding the interactions between objects and the analysis of motion.

Teaching tips

Exam questions on forces generally fall into two categories – questions involving identifying forces, classifying them as contact or non-contact, and calculating resultant forces; and questions involving applications of Newton's Laws of motion. These are frequently combined together.

Newton's First Law of motion involving an object, such as a parachutist falling at terminal velocity, is a great example to use with students, as it combines many aspects of this topic, such as identifying the forces of gravity and air resistance, Newton's Laws of motion and calculating resultant forces.

Forces questions are frequently used to test students' understanding of converting unit prefixes into numbers that can be used in calculations, with kN being the classic example at foundation level.

The two required practicals associated with this topic (Investigating acceleration and Investigating the behaviour of springs) are commonly used in exam questions.

Other common contexts include different sports; moving vehicles; stopping distances; or other falling objects. You can help students to feel more confident about this topic by giving them plenty of practice of calculating the resultant forces acting on moving objects, and then applying Newton's Second Law of motion.

SOLVE

Now turn back to the original question with the student. Work through the writing frame in the workbook together. The solutions are given below.

Forces can affect the motion of objects.

The badminton shuttlecock shown below is falling downwards at constant speed.

a) State the law that gives the relationship between the two forces acting on the shuttlecock when it is moving at a constant speed.

b) Draw arrows on the diagram to show the two forces acting on the shuttlecock.

c) Describe what happens when the shuttlecock hits a stationary badminton racket facing upwards. You can use a diagram to help you explain.

For part a): Which one of Newton's laws does this show? Decide which one deals with equal and opposite reactions.

Newton's Third Law.

For part b): If the shuttlecock is moving at a steady speed then the forces must be balanced. What are the two main forces? Add arrows to the diagram of the shuttlecock to show these forces.

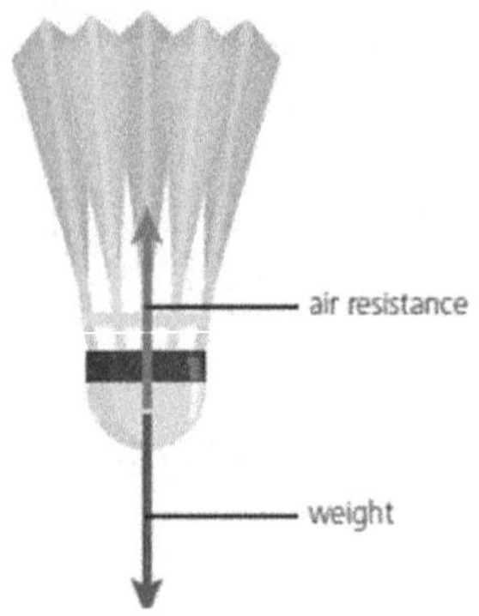

For part c): Try drawing the shuttlecock as it makes contact with the racket. Now add arrows to your diagram to show the two forces. How will you use the size of your arrows to show the size of the forces?

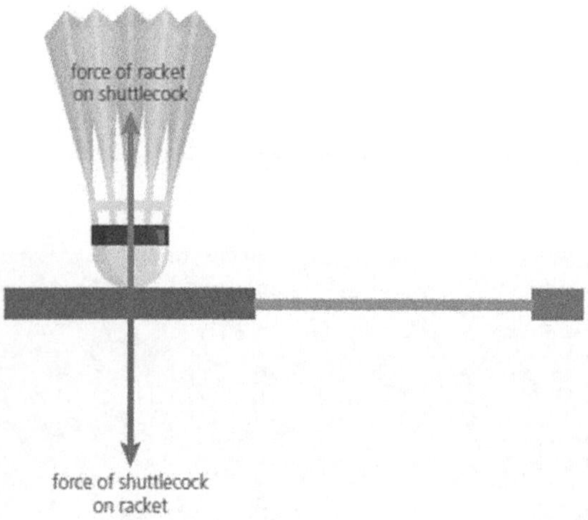

For part c): Describe what happens to the shuttlecock as it makes contact with the racket.

The shuttlecock exerts a downwards force on the racket. The racket exerts an equal and opposite force on the shuttlecock acting upwards.

The shuttlecock rebounds/moves off the racket upwards.

PRACTISE

Use the additional practice questions on this powerful idea at the back of the workbook in future sessions. Additional practice questions for Session 23 can be found in the Estimate section.

SESSION 24: MOTION

THINK

Ask the student to read the question for Session 24 in the workbook and scribble down what they know on that topic. Now ask the student to have a go at the questions on Adapt© for Session 24.

Common misconceptions

Some students think that velocity is another word for speed, and that they mean the same thing.	To address this, explain that speed is a scalar quantity, and only has size; velocity is a vector quantity, and has size and direction – velocity is the speed of an object in a given direction. See key idea cards P24.2 and P24.4.
Some students think that horizontal line on a velocity–time graph shows a stationary object.	To address this, explain that this is true if the horizontal line is at 0 m/s, but otherwise it represents an object moving at constant velocity. See key idea card P24.7.
Many students think that vehicles with more mass have longer braking distances.	To address this, explain that, although larger mass vehicles exert more force on the road surface, they produce more friction, and so stop in the same distance – more massive vehicles also have bigger brakes that exert more force to stop the wheels. See key idea card P24.3.
Many students think that when a parachute opens, a parachutist moves upwards.	To address this, explain that the parachute increases the upwards force, so that there is a resultant force acting upwards, which causes a deceleration, so the parachutist slows down, but does not move upwards. See key idea card P23.5.

CONNECT

The student should watch the video on the powerful idea for the session. The powerful idea for Session 24 is that scalar quantities, such as mass, only have size (or magnitude). Quantities such as force, have size and direction and are called vector quantities. The motion of objects can be described in terms of distances, speeds, velocities and accelerations. The best way to show the motion of an object is by using a distance–time graph or a velocity–time graph.

Knowledge of the scalar/vector nature of quantities helps with the understanding of other key concepts in physics such as electricity, energy and electromagnetism. Distance/time and velocity/time graphs are very common examination questions allowing examiners to test students' understanding of graphical analysis and calculation.

Teaching tips

Exam questions on motion focus heavily on the analysis of graphs of motion, and then calculations of quantities such as velocities and accelerations. Students can be given distance–time graphs or velocity–time graphs and be asked to describe the motion or to calculate velocity/speed from a distance–time graph and acceleration from a velocity–time graph. (Note: Calculating distance travelled from the area under a velocity–time graph is higher tier only).

Students are frequently asked to identify units for quantities such as velocities and acceleration. Questions on motion are also commonly combined with the concept of work done (from the energy topic), with students being asked to calculate the work done when a force moves an object through a given distance.

The required practical (Investigating acceleration) could be used in questions on this topic, and such questions would mostly be linked to Session 23: Forces, and Newton's Laws of motion, particularly Newton's Second Law of motion.

Most of the common contexts involve the motion of objects such as cars, cyclists and runners. Sometimes, students may be asked to plot graphs from tables of data or sketch motion graphs. You can help students to feel more confident about this topic by giving them plenty of practice of drawing, interpreting and making calculations from distance–time graphs and velocity–time graphs.

SOLVE

Now turn back to the original question with the student. Work through the writing frame in the workbook together. The solutions are given below.

The graph shows a velocity–time graph for a girl riding a bicycle from point A to point E.

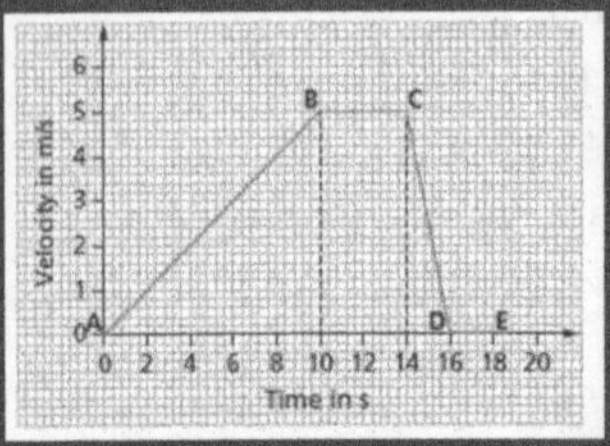

a) Describe, in words, the girl's journey from A to E.

b) i) Determine the change in velocity from A to B.

ii) Calculate the acceleration of the girl from A to B.

c) Calculate the distance that the girl travels between B and C.

Use the equation:

distance (m) = velocity (m/s) × time (s)

To answer part a), break the graph down into steps. In the space below, write what is happening between the labelled points on the graph. The question asks you to use words only so there is no need to write any numbers here.

The girl's journey from A to E is summarised as:

- A to B: moving with constant acceleration
- B to C: moving at constant velocity
- C to D: moving with a constant deceleration (value higher than the initial acceleration)
- D to E: stationary.

For part b) i) you first need to read values from the graph. Look at the y-axis value after 10 seconds. How much has this value changed from 0 seconds?

change of velocity = 5 m/s – 0 m/s = 5 m/s

For part b) ii), you need to calculate the acceleration from A to B. Put the values for change in velocity and time into the equation: acceleration (m/s²) = change in velocity (m/s) / time taken (s). Remember to write the units.

$$\text{Acceleration} = \frac{\text{change in velocity}}{\text{time taken}}$$

$$= \frac{5\ \text{m/s}}{10\ \text{s}} = 0.5\ \text{m/s}^2$$

For part c), what is the constant velocity between B and C? Then calculate the time between B (10 s) and C (14 s). Put these two values into the equation given:

distance (m) = velocity (m/s) × time (s)

What will the units be?

Between B (10 s) and C (14 s) the girl travels for 4 s.

$$\text{distance} = \text{velocity} \times \text{time}$$

$$= 5\ \text{m/s} \times 4\ \text{s} = 20\ \text{m}$$

PRACTISE

Use the additional practice questions on this powerful idea at the back of the workbook in future sessions. Additional practice questions for Session 24 can be found in the Calculate, Explain and Plot sections.

SESSION 25: WAVES

THINK

Ask the student to read the question for Session 25 in the workbook and scribble down what they know on that topic. Now ask the student to have a go at the questions on Adapt© for Session 25.

Common misconceptions

Many students think that sound can travel through space (a vacuum).	To address this, explain that sound is a longitudinal wave and travels by the successive vibration of particles from a source to a detector. Space is a vacuum. There are no particles. So, in space, no one can hear you scream! See key idea card P25.2.
Many students think that the amplitude of a wave is the distance from the bottom of a trough to the top of a crest.	To address this, explain that the amplitude of a wave is the difference between the maximum displacement of the wave (e.g. a crest) and the undisturbed position of the wave – this can be demonstrated using a slinky. See key idea cards P25.2 and P25.5.
Many students think that all waves need a medium (particles) in order to travel through something.	To address this, explain that light does not need particles in order to travel through something – we can see the light that comes from the stars, even though it travels through a vacuum. See key idea card P25.3.
Many students think that the electromagnetic spectrum is a set of monochromatic regions.	To address this, explain that the electromagnetic spectrum is a continuous 'spectrum' of wavelengths – for example, mobile phones use microwaves with a wavelength of about 30 cm and microwave ovens use microwaves with a wavelength of 12 cm. See key idea card P25.3.

CONNECT

The student should watch the video on the powerful idea for the session. The powerful idea for Session 25 is that waves transfer energy. Longitudinal waves (e.g. sound waves) cause the particles of the substance the wave is travelling through to move backwards and forwards. Transverse waves (e.g.water waves) make the particles move at a right angle to the direction the wave is travelling. Different parts of waves can be measured to understand the properties of waves.

Students need to have a good understanding of the key terms used to describe the nature and behaviour of waves. Using and understanding the correct waves vocabulary substantially enhances students' abilities to deconstruct the meaning of examination questions and successfully answer them.

Teaching tips

Exam questions on waves generally fall into two categories. First, students get asked questions about the basic properties of transverse and longitudinal waves, such as identifying and calculating wavelengths, frequencies, periods, amplitudes and speeds. Second, questions ask about the properties and uses of the electromagnetic spectrum. It is common to see questions about identifying missing parts of the spectrum and also the use of the wave equation:
speed = frequency × wavelength.

The electromagnetic spectrum is a common context for testing student's understanding of unit prefixes and their conversion. The wave equation is also used as the basis for testing students' abilities to rearrange formulae.

The required practical (Investigating waves in a ripple tank and waves in a solid (on a string) is often used as a context for the delivery of questions involving the basic properties of waves, and students are frequently asked to calculate the speed of water waves from ripple tank measurements. Waves on a string are typically used as a context for annotating different features of a transverse wave, such as wavelengths and amplitudes.

The most common contexts for the electromagnetic spectrum involve communications, particularly mobile phones, and the medical use of ionising radiation. You can help students to feel more confident about this topic by giving them practice of annotating transverse waves, testing their ability to recall the parts of the electromagnetic spectrum, in order, and doing calculations involving the wave equation.

SOLVE

Now turn back to the original question with the student. Work through the writing frame in the workbook together. The solutions are given below.

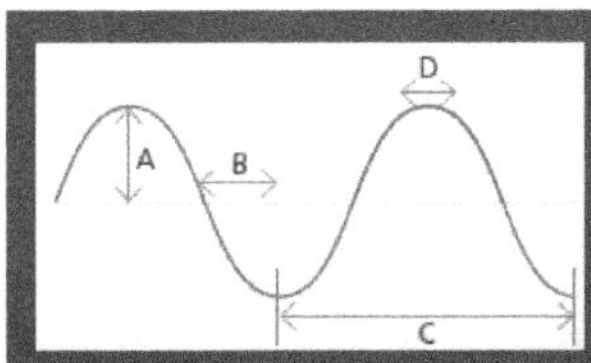

The diagram shows a transverse water wave. The frequency of the wave is 3 Hz and the wavelength of the wave is 2 m.

a) Identify the label corresponding to the:

i) amplitude

ii) wavelength.

b) Calculate the speed of the wave using the wave equation.

c) The table below shows the parts of the electromagnetic spectrum. Some of the parts are missing.

radio waves		infrared		ultraviolet		gamma rays

i) Write the missing words in the table.

ii) Draw an arrow under the table to show in which direction wavelength increases.

iii) Explain why ultraviolet rays are dangerous to humans.

For part a): What are the characteristics of a transverse (e.g. water) wave? Where do we measure amplitude and wavelength from?

i) Amplitude = A

ii) Wavelength = C

For part b): Gather all the numerical information from the question that you need to use the wave equation:

wave speed (m/s) = frequency (Hz) × wavelength (m)

Then substitute the numbers into the equation and use a calculator to work out the answer. Repeat this on your calculator to check that you have entered the numbers correctly. What are the units of wave speed in this question?

wave speed = frequency × wavelength

= 3 Hz × 2 m = 6 m/s

For part c) i), ii): List all of the parts of the electromagnetic spectrum in the correct order. Try to remember which have the longest wavelength, and so which way the arrow should point.

i)

radio waves	**microwaves**	infrared	**visible light**	ultraviolet	**X-rays**	gamma rays

ii) A horizontal arrow pointing to the left should be drawn under the table.

For part c) iii): You need to ***explain*** *in this question – this means you need to state the reasons for something. First, state the property of ultraviolet rays that makes them dangerous. Next, give the reasons why this makes them dangerous.*

Ultraviolet rays are dangerous to humans because they have a high energy and can harm or kill cells.

PRACTISE

Use the additional practice questions on this powerful idea at the back of the workbook in future sessions. Additional practice questions for Session 25 can be found in the Estimate and Show sections.

SESSION 26: ENERGY

THINK

Ask the student to read the question for Session 26 in the workbook and scribble down what they know on that topic. Now ask the student to have a go at the questions on Adapt© for Session 26.

Common misconceptions

Many students think that energy is created by systems and can be used up.	To address this, explain that energy is stored in systems and is transferred from store to store – a simple demonstration of a moving pendulum shows the energy transferring repeatedly from gravitational potential energy to kinetic energy. See key idea card P26.6.
Some students think that energy is some form of 'intangible fluid' that flows from one place to another during a process.	To address this, explain that energy is stored in a variety of different forms and can transfer from one store to other stores – demonstrate a falling object such as a pen and discuss the energy transfers that occur. See key idea cards P26.2 and P26.3.
Many students think that appliances labelled as being efficient are always 100% efficient.	To address this, explain that efficiency is frequently used as a labelling and marketing tool to sell appliances – most efficiency rating systems, such as the letter system, are really a comparison system, of comparisons with the least efficient appliances, and do not directly relate to percentage efficiency. See key idea card P26.4.
Many students think that energy is another word for fuel.	To address this, explain that the ability of a fuel to do a job, such as petrol in a car engine, is because it has a store of energy associated with it (chemical in this case). See key idea card P26.5.

CONNECT

The student should watch the video on the powerful idea for the session. The powerful idea for Session 26 is that energy stores are needed for actions. Energy is transferred from one store to another during an action. A force causes the transfer of energy from one store to another. For example, when a ball is dropped the gravitational potential energy stored in the ball is transferred to kinetic energy. The rate of transfer of energy is called power. More powerful devices can transfer more energy per second.

Knowledge of energy stores and transfers is fundamental to an understanding of physics. This concept underpins most of the other ideas in physics. It is worth investing time in ensuring that students have a clear understanding of the difference between an energy store and an energy transfer process before starting this topic.

Teaching tips

Exam questions on energy generally fall into two categories. The first category involves energy transfers, and questions typically include calculations involving the equations for gravitational potential energy, kinetic energy and power. The second category involves the generation of electricity via renewable and non-renewable sources, and questions frequently ask students to compare one source of generation to another.

A common theme that runs through both categories is the concept of energy efficiency, and students are frequently asked to calculate efficiencies and to evaluate methods of energy efficiency, particularly in the context of home heating.

The required practicals (Investigating specific heat capacity, and Investigating infrared radiation) are frequently used as contexts for questions involving energy. The specific heat capacity equation is often set as a question, with students being asked to calculate energies or specific heat capacities (although the rearranged equation is often given for this) from data given in the question. The infrared radiation experiment is sometimes combined with questions about the electromagnetic spectrum, but is usually associated with thermal transfer.

A wide range of contexts are used for energy questions, although fairground rides, domestic appliances, home insulation and power stations are quite common. You can help students to feel more confident about this topic by giving them practice of recalling and using the equations involved.

SOLVE

Now turn back to the original question with the student. Work through the writing frame in the workbook together. The solutions are given below.

A man, with a mass of 75 kg, climbs a flight of steps 1.5 m high.

At the top of the steps he stops, before jumping off the top step and landing back down on the ground.

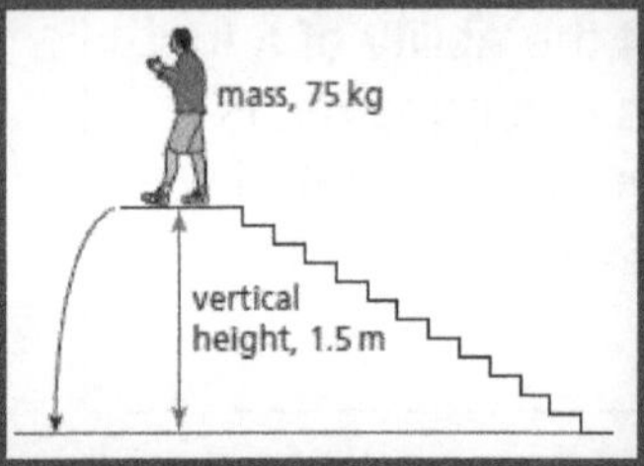

a) Describe the changes in energy as he climbs the steps, stops, and then jumps down.

b) Calculate the gravitational potential energy of the man at the top of the steps.

The gravitational field strength is 10 N/kg.

c) The velocity of the man, just before he hits the floor is 5.2 m/s.

Calculate the kinetic energy of the man just before he hits the floor.

d) Use your answers to parts b) and c) to calculate the energy wasted as sound energy and thermal energy as the man falls.

For part a), you need to describe how the energy transfers from kinetic energy into gravitational potential energy and back again. How is energy wasted?

Kinetic energy of the man moving up the steps is transferred to gravitational potential energy at the top of the steps, with some wasted as sound and thermal energy. This gravitational potential energy is then converted back into kinetic energy as the man falls, with some energy wasted as sound and thermal energy.

For part b), you need to recall the equation for gravitational potential energy:

Gravitational potential energy = mass × gravitational field strength × height

Underline all the values you have been given in the question and/or on the diagram. Now insert them into the equation. Can you remember the units? Use a calculator to work out the answer. Remember to check your answer by repeating the calculation on your calculator.

Gravitational potential energy = mass × gravitational field strength × height

= 75 kg × 10 N/kg × 1.5 m

= 1125 J

For part c), you need to recall the equation for kinetic energy:

kinetic energy = 0.5 × mass × velocity^2

Underline all the values you have been given in the question and/or on the diagram. Now insert them into the equation. Can you remember the units? Use a calculator to work out the answer. Remember to check this by repeating the calculation on your calculator.

kinetic energy = 0.5 × mass × velocity^2

= 0.5 × 75 kg × (5.2 m/s)^2

= 1014 J

For part d), remember that the law of conservation of energy states that energy does not just disappear. Therefore, the gravitational potential energy at the top of the steps is transferred into kinetic energy plus wasted energy.

energy wasted = gravitational potential energy – kinetic energy

Use the values from parts b) and c) in your calculation. Remember to write the units. Now use a calculator to work out the answer. Remember to check this by repeating the calculation on your calculator.

All the gravitational potential energy at the top of the steps will be transferred into kinetic energy plus the wasted energy, so:

energy wasted = gravitational potential energy – kinetic energy

= 1125 J – 1014 J

= 111 J

PRACTISE

Use the additional practice questions on this powerful idea at the back of the workbook in future sessions. Additional practice questions for Session 26 can be found in the Compare and Describe sections.

SESSION 27: ELECTRICITY

THINK

Ask the student to read the question for Session 27 in the workbook and scribble down what they know on that topic. Now ask the student to have a go at the questions on Adapt© for Session 27.

Common misconceptions

Many students think that electric current, rather than energy, is stored in a battery.	To address this, explain that current is the flow of charge around a circuit – a simple analogy of a pump moving water around a closed series of pipes (e.g. a central heating system) should clear this up. See key idea card P27.1.
Some students think that two resistors in parallel will have a larger resistance than each resistor alone.	To address this, explain that adding a second resistor to either of the resistors on their own, provides an extra path for current to flow, so more current flows in a circuit with two resistors in parallel. See key idea card P27.5.
Many students think that as wires get longer, the current finds it more and more difficult to pass through.	To address this, explain that the wire contains conducting, 'free' electrons along its whole length and the same current flows through each part of the wire. See key idea card P27.3.
Many students think that mains sockets have current inside them, waiting to get out, and child socket covers stop it leaking out.	To address this, explain that a mains electric socket provides a way for energy to transfer to an external circuit and be used to create an electric current, which will only flow if the external circuit is complete. See key idea card P27.4.

CONNECT

The student should watch the video on the powerful idea for the session. The powerful idea for Session 27 is that an electrical circuit is a closed loop of conducting material with at least one cell and a component (e.g. a lamp). The cell is the source of electrical energy. Electric current is the rate of flow of charge through the circuit. As current flows, it is opposed by the structure of the circuit causing resistance. In a series circuit, the current is the same all the way round the circuit. A parallel circuit has different loops for the current to flow through. The total current of all the loops in a parallel circuit is equal to the current flowing through the cell.

Students benefit greatly in this session from learning the different circuit symbols and remembering the key equations. Examiners frequently set questions involving the analysis of an electrical circuit, drawing or interpreting an IV graph, and performing calculations to determine unknown quantities.

Teaching tips

Exam questions on electricity generally involve analysing series and parallel circuits, the electrical characteristics of components, and mains electricity. All three types of question can involve the use of the equations for charge, Ohm's Law and power. Students are often asked to calculate the overall resistance of circuits and interpret electrical characteristic (*IV*) graphs for a range of different components. The three-pin mains plug, cable colours and the differences between a.c. and d.c. are common questions.

The electricity topic includes many equations that students need to recall as they are not provided in the exam. Teachers can help students remember the equations by practicing recalling them from memory. Students sitting foundation tier are expected to calculate resistances in series, but in parallel circuits they only need to know that the total resistance of two resistors in parallel will be less than the lowest resistance.

The required practicals (Investigating *IV* characteristics and Using circuit diagrams) are frequently asked about in exam questions on electricity. Common questions include plotting graphs of resistance against the length of wires and *IV* characteristics.

You can help students to feel more confident about this topic by giving them lots of practice of plotting and interpreting graphs of resistance against length and *IV* characteristics for a range of different components.

SOLVE

Now turn back to the original question with the student. Work through the writing frame in the workbook together. The solutions are given below.

A student performs an experiment to measure the potential difference across, and current through, a fixed resistor. She collects the following results:

Potential difference in V	0	2	4	6	8	10
Current in A	0.0	0.1	0.2	0.5	0.4	0.5

a) Plot a graph of these results. Plot the potential difference on the *x*-axis and current on the *y*-axis.

b) One of the results has been recorded incorrectly and is an anomaly. Identify the anomaly and draw a circle around this result on your graph.

c) Draw a line of best fit through the valid data points.

d) Suggest a correct value for the anomalous result.

e) The equation for resistance is:

$$\text{resistance} = \frac{\text{potential difference}}{\text{current}}$$

Calculate the resistance of the fixed resistor.

Use the result for a potential difference of 8 V and the equation above.

*For part a) you need to **plot** a graph. First, look at the data given. You are told which row needs to be plotted on the x-axis and which one on the y-axis. Does the graph start at (0, 0)? What are the maximum x-axis and y-axis values? Draw a scale that fits both using a pencil and a ruler (in case you make a mistake). Plot the data points accurately using the scales. Remember to label your axes using the table row headings.*

*For part b), can you see an **anomaly** (a data point that does not fit the pattern)? Circle it on your graph.*

For part c), think about lines of best fit. Is this one a straight line, or a curve? Does it go through (0, 0)? Do you include the anomaly?

a) to c)

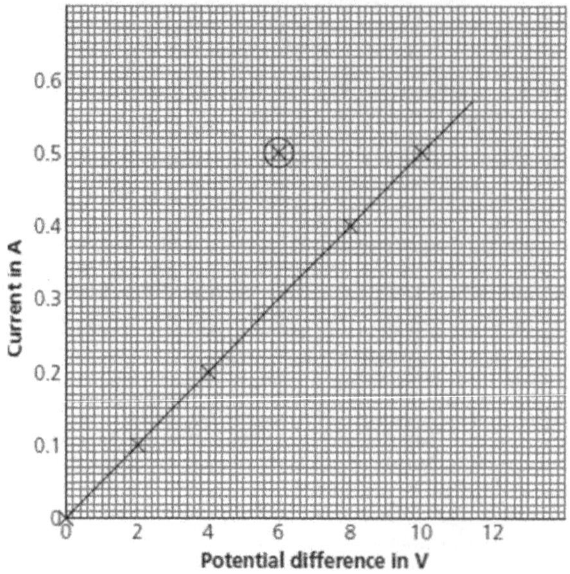

For part d), now use the line of best fit that you have drawn to identify the true value of the current at 6 V.

potential difference = 6 V; current = 0.3 A

For part e), read the value of the current off the graph corresponding to a potential difference of 8 V. Write down the data values that you have read off the graph, and the equation given. Insert the values into the correct place in the equation and use a calculator to calculate your answer. Repeat the calculation using your calculator to double check your answer. Write down your final answer and write down the unit of resistance. You can use the symbol or the word.

At a potential difference of 8 V, current = 0.4 A, so:

$$\text{Resistance} = \frac{\text{potential difference}}{\text{current}}$$

$$\text{Resistance} = \frac{8\text{ V}}{0.4\text{ A}} = 20\ \Omega\text{ (or 20 ohms)}$$

PRACTISE

Use the additional practice questions on this powerful idea at the back of the workbook in future sessions. Additional practice questions for Session 27 can be found in the Define and Draw sections.

SESSION 28: ELECTROMAGNETISM

THINK

Ask the student to read the question for Session 28 in the workbook and scribble down what they know on that topic. Now ask the student to have a go at the questions on Adapt© for Session 28.

Common misconceptions

Many students think that you can separate a bar magnet into its north pole and its south pole.	To address this, explain that cutting a magnet in half would only produce two magnets, each with a north and south pole – this can be demonstrated with several magnets in a line forming a giant long magnet. See key idea card P28.2.
Some students think that a magnetic field is a pattern of lines surrounding a magnet.	To address this, explain that a magnetic field is a place where a magnetic material experiences a force, the field exists in between the field lines as well. See key idea card P28.5.
Many students think that electromagnets are made from a magnet inside a coil of wire.	To address this, explain that when current flows through a metal wire, it generates a magnetic field. A permanent magnet always has a magnetic field associated with it. The magnetic field of an electromagnet can be turned on and off. See key idea cards P28.3 and P28.4.
Some students think that current flowing through a single wire does not produce a magnetic field.	To address this, explain that an electromagnet is a coil of wire with current flowing through it. The magnetic field around a coil does decrease as the number of coils is reduced, but it does not drop to zero when there is just a single wire. See key idea card P28.3.

CONNECT

The student should watch the video on the powerful idea for the session. The powerful idea for Session 28 is that magnetism is a non-contact force between two poles of a magnet. The magnetic fields around magnets can be drawn using magnetic field lines. The magnetic field is strongest at the poles of a magnet. Electromagnets are formed when an electric current flows through a conductor such as a metal wire. The shape of the magnetic field around an electromagnet depends on the shape of the electromagnet.

Students should be able to: plot the magnetic fields around permanent magnets and electromagnets; be able to analyse the interactions between magnets and non-magnets; and explain the action of an electromagnet.

Teaching tips

Exam questions on electromagnetism tend to fall into two categories: questions on permanent magnetism, and questions on electromagnets. Common questions on permanent magnets focus on the shapes of magnetic fields, the rules for pole interactions, and induced magnetism. Questions on electromagnetism concentrate on the shapes of fields around different configurations of electromagnets, and the results of experiments investigating the strength of electromagnetic fields.

The electromagnetism topic is often used to ask students to plan/design experiments, or to plot/analyse graphs involving the strength of a magnetic field. Drawing magnetic field lines is also a common theme for these questions – remember to emphasise to students that:

- magnetic field lines have a direction
- arrows should be drawn on field lines that point from north poles to south poles
- the space between the lines is important – the closer the lines are, the stronger the magnetic field is.

You can help students to feel more confident about this topic by giving them practice of drawing magnetic field lines around bar magnets, single wires and solenoids. They should also practise plotting graphs involving the strength of magnetic fields, such as plotting the number of paper clips that can be supported by an electromagnet with different currents running through it, or with different numbers of coils.

SOLVE

Now turn back to the original question with the student. Work through the writing frame in the workbook together. The solutions are given below.

A child's toy magnet set, containing three solid metal rods, has become mixed up.

• One rod is a permanent magnet.

• The second rod is made from iron.

• The third rod is made from aluminium.

All the rods look the same.

Explain how a permanent bar magnet could be used to determine the nature of each rod.

What are the rules for the interaction of magnetic poles? How could these be used to identify the permanent magnet rod?

Like poles repel and unlike poles attract.

Bring each pole of the permanent bar magnet up to each rod in turn.

If the end of the rod is attracted to one pole of the permanent magnet, but it is repelled by the other pole, then the rod is a permanent magnet.

What is induced magnetism? What materials can be made into an induced magnet? Does it matter which pole of the permanent magnet is used to induce magnetism? How could this be used to identify the iron rod?

Induced magnetism occurs when a permanent magnet is brought up next to an object made from iron, steel, cobalt or nickel. The permanent magnet 'induces' (creates) a magnetic field in the object, making it behave like another permanent magnet. Both poles of the permanent magnet induce an opposite pole in the object next to it. It doesn't matter which way around the object is. If both ends of the rod are attracted to both poles of the permanent magnet, then the rod is made from iron, because iron can be made into an induced magnet with either of the ends of the permanent magnet.

What happens when a permanent magnet is brought up close to a material that cannot be magnetised? How could this be used to identify the aluminium rod?

If one end of the rod is not attracted to either pole of the permanent magnet, then the rod is made from aluminium, because aluminium cannot be magnetised.

Now bring together all this information to explain how to determine the nature of each rod.

The rod that is a permanent magnet is attracted to one pole of the permanent magnet but is repelled by the other pole.

The iron rod is attracted to both poles of the permanent magnet.

The aluminium rod is not attracted to either pole of the permanent magnet.

PRACTISE

Use the additional practice questions on this powerful idea at the back of the workbook in future sessions. Additional practice questions for Session 28 can be found in the Predict section.

SESSION 29: ATOMS AND RADIOACTIVITY

THINK

Ask the student to read the question for Session 29 in the workbook and scribble down what they know on that topic. Now ask the student to have a go at the questions on Adapt© for Session 29.

Common misconceptions

Many students think that isotopes are different elements because their nuclei are different.	To address this, explain that isotopes of elements have different numbers of neutrons, but they all have the same number of protons (the same atomic number). See key idea cards P29.1 and P29.6.
Many students think that there is air between the subatomic particles in atoms.	To address this, explain that air is made up of atoms, and an atom cannot have other atoms inside it – the space inside an atom is empty space. See key idea cards C13.2 and P29.2.
Many students think that beta particles are electrons so they come from the atom's electron shells.	To address this, explain that beta particles are produced when a neutron (in the nucleus), decays into a proton and an electron. See key idea card P29.3.
Students think that during beta decay the atomic number of the daughter nucleus reduces by one.	To address this, explain that during beta decay, a neutron decays into a proton and an electron. The mass number must stay the same, but the atomic number must go up by one. See key idea card P29.3.
Many students think that half-life is the time for radioactivity to disappear.	To address this, explain that half-life is the time taken for half the number of radioactive atoms in a substance to decay, so the substance remains radioactive after one half-life. See key idea card P29.4.

CONNECT

The student should watch the video on the powerful idea for the session. The powerful idea for Session 29 is that atoms are the tiny, basic building blocks of matter. The nuclear model of the atom is a small nucleus containing protons and neutrons, surrounded by orbiting electrons. Isotopes of atoms have the same number of protons but different numbers of neutrons. Some atoms are unstable and emit nuclear radiation in a random process called radioactive decay. Nuclear radiation is hazardous to living cells and precautions need to be taken when handling radioactive substances.

This session crosses over with the chemistry session 13: Atomic structure. The emphasis in this session is on the nature of isotopes and the mechanisms, patterns and hazards of radioactive decay.

Teaching tips

Exam questions about atoms and radioactivity fall into three general categories:

- isotopes, the AZX notation, nuclear decay equations and the development of the atomic/nuclear model
- the nature and properties of the three main types of nuclear radiation (α, β, γ)
- the patterns involved with radioactive decay, and the determination of half-life.

This topic often involves balancing a nuclear decay equation and then determining the number of protons and neutrons in an isotope. It also gives opportunities to compare the nature and/or properties of α, β, γ radiation, frequently by asking students to compare the absorption of these radiations by different materials. It is common for students to be given data about radioactive decay, and then be asked to plot decay graphs. They may be asked to analyse a decay graph, and determine the half-life using a graphical technique such as drawing lines on the decay graph to show and measure the half-life.

You can help students to feel more confident about this topic by giving them practice of balancing nuclear decay equations and calculating the number of protons and neutrons in a wide number of different isotopes – include plenty of practice involving beta decay. You should also give students practice of determining the half-life of isotopes using a graphical technique from decay graphs.

SOLVE

Now turn back to the original question with the student. Work through the writing frame in the workbook together. The solutions are given below.

The table below shows how the activity of a radioactive sample of lead-217 (atomic number 82) changes over time.

Time in s	0	10	20	30	40	50
Activity in Bq	120	85	60	42	30	21

a) Plot a graph of activity against time.

b) Draw a line of best fit through the points.

c) Use information from the graph to determine the half-life of lead-217.

d) Lead-217 decays by beta particle emission into an isotope of bismuth, Bi. Balance the nuclear decay equation for this decay:

$$^{217}_{82}\mathrm{Pb} \rightarrow {}^{0}_{-1}\mathrm{e} + {}^{\cdots}_{\cdots}\mathrm{Bi}$$

For part a): You will first need to determine the axes for the decay graph. Which quantity goes on which axis? What scale do you need – what are the lowest and the highest values that you need to plot on each axis? Do you need to start the graph axes at (0, 0)? Make sure that you plot the points accurately using a sharp pencil. *Double-check them.*

For part b): Think about the general shape plotted out by the points (is it a curve or a straight line)? Remember, a line of best fit does not need to go through all the points, but it does need to be a smooth line, not join-the-dots.

a) and b)

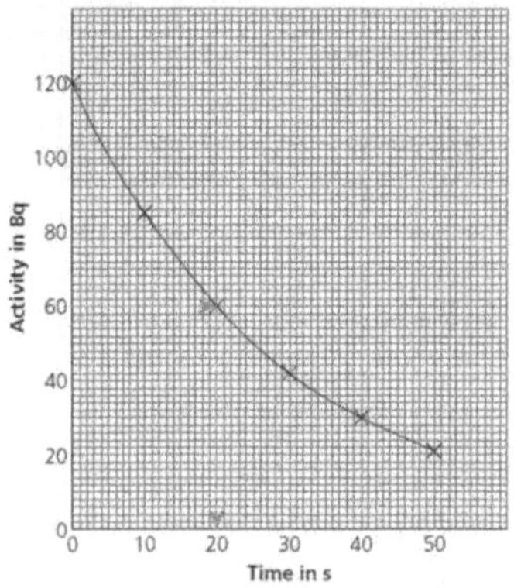

For part c): The half-life is the time taken for the initial activity to fall to half its value. What is the initial activity? What is half of this initial value? Draw a straight line across the graph from this value, so it touches the line of best fit. Then draw a line down to the time axis and read off the half-life. What are the units?

The horizontal arrow shows half the initial activity; this hits the line of best fit at a time of 20 s. This is the half-life.

For part d): You need to balance the nuclear decay equation. The top numbers, mass numbers, need to be the same on each side of the equation. What is the mass number of the bismuth isotope? The bottom numbers, atomic numbers, also need to balance on either side of the equation. Remember the value for a beta particle is −1, so the atomic number of bismuth – 1 = atomic number of lead. What is the atomic number of the bismuth isotope?

$$^{217}_{82}\text{Pb} \rightarrow \, ^{0}_{-1}\text{e} + \, ^{217}_{83}\text{Bi}$$

The missing mass number is 217; the missing atomic number is 83.

PRACTISE

Use the additional practice questions on this powerful idea at the back of the workbook in future sessions. Additional practice questions for Session 29 can be found in the Compare, Define, Determine and Justify sections.

SESSION 30: PHYSICS PRACTICALS

THINK

Ask the student to read the question for Session 30 in the workbook and scribble down what they know on that topic. Now ask the student to have a go at the questions on Adapt© for Session 30.

Common misconceptions

Many students think that ice cannot change temperature and is always at 0 °C.	To address this, explain that ice inside a freezer will have the same temperature as the inside of the freezer and its temperature increases when it is removed.
Many students think that the gradient of an *IV* graph is the resistance of the component.	To address this, explain that resistance is given by the equation: $R = V/I$ at any point along the *IV* graph. See key idea cards P27.3 and P30.5.
Many students think that the current gets less as it passes through a component.	To address this, explain that the current in a circuit is dictated by the total resistance of a circuit. The current is the same before and after a component, and is the same around a series circuit. See key idea cards P27.1 and 27.2.
Students think that heavier objects are denser.	To address this, explain that density depends on weight and volume; a large ship will be very heavy, but it occupies a large volume, so its overall density can be lower than that of water, and so it floats. See key idea cards C12.3 and P30.3.
Many students think that speed and acceleration mean the same thing.	To address this, explain that speed is a measure of how fast or slow an object is moving, but acceleration is the rate at which it is changing speed – a sports car can reach 60 mph in a much shorter time than a large truck. See key idea cards P24.2 and P30.4.
Many students think that the extension of a spring is the same as the length of the spring.	To address this, explain that the extension of a spring is the extra length gained when it is stretched – this can be easily demonstrated with a standard school spring. See key idea card P30.6.
Many students think that the standing waves formed on a vibrating string are not moving.	To address this, explain that, although the peaks and troughs do not move from one place to another, the particles of the string are still vibrating up and down – placing a finger lightly on one of the peaks will allow the student to feel the vibration. See key idea card P30.8.

Some students think that the different sides of a Leslie cube must have different temperatures because some feel hotter than the others.	To address this, explain that the Leslie cube is made of copper, which is in contact with the hot water inside. Each metal side will have the same temperature. Differences in the heating effect of each side are due to the colour and nature of the surface and affect the emission of infrared radiation. See key idea card P30.13.
Many students think that newtonmeters can be used without checking for zero error.	To address this, explain that newtonmeters must be checked for zero error before they are used, otherwise the measurements taken will not be valid. See key idea card P30.9.
Many students think that the speed of an object rolling down a ramp can be determined by measuring the length of the slope and timing how long it takes to roll down the slope.	To address this, explain that as an object rolls down a slope it accelerates (speeds up), so the measurement of speed calculated this way will be a mean average speed, not the instantaneous speed at any point down the slope. See key idea card P30.9.
Many students think that the gradient (slope) of an inclined straight line on a graph can be determined by 'counting squares' vertically and horizontally and dividing the vertical squares by the horizontal squares.	To address this, explain that the vertical displacement is determined using the *y*-axis scale and the horizontal displacement by the *x*-axis scale. See key idea card P30.10.
Many students think that a safety measure for every physics practical experiment is to wear a pair of safety spectacles.	To address this, explain that safety spectacles are only appropriate if the identified hazards can harm the eyes. If they cannot, then safety spectacles do not need to be worn. See key idea card P30.11.
Many students think that in an examination you will always be given any equations that you need to use.	To address this, explain that each examination will have a designated set of equations that you need to learn. The equations may not be given, and students may have to recall them from memory, so they need to learn them. See key idea card P30.11.

CONNECT

In this session students need to familiarise themselves with the practical layout and methods of the eight physics required practicals; their associated 'working scientifically' skills and techniques; and to have an awareness of some of the key physics skills needed throughout the course and for success in the examination.

Teaching tips

Exam questions about the required practicals are used as mechanisms to ask questions about the range of practical skills outlined in the specification under 'Working scientifically – Experimental skills and investigations'.

Questions involving the investigation of specific heat capacity often focus on instruments and apparatus. Students may be asked to produce a list of apparatus or draw a diagram/circuit diagram of an experiment that could be used to determine the specific heat capacity of a material. They may then be asked to identify the key measurements that are needed to determine a value. You can help students to feel more confident about this practical by making sure that they can recall and use the equation involving specific heat capacity, and then assigning ways to record the different quantities needed.

Questions about investigating *IV* characteristics often involve analysing data given either as a table or in graphical format. It is common for students to be asked to plot an *IV* graph and determine a resistance from it. This experiment also gives examiners opportunities to test students' abilities to identify anomalous results and calculate means. You can help students to feel more confident about this practical by making sure that they can plot *IV* graphs confidently, and that they know how to apply the Ohm's Law equation.

The practical using circuit diagrams offers opportunities to draw circuit diagrams using the correct circuit symbols, and to plan/design an experiment to investigate the variation of resistance with length for a wire. In this case, students should be prepared to create a hypothesis and identify the variables involved in this investigation. They may also be asked to write a method and identify ways to minimise uncertainties. You can help students to feel more confident about this practical by making sure that they can identify and draw circuit symbols correctly, and that they know what a hypothesis is and how to identify independent, dependent and control variables.

The practical investigating density is frequently set as a way of asking students to describe a method for determining the density of a regular or irregular object. Students often forget to measure the mass of the object, as they focus on ways to measure volume. You can help students to feel more confident about this practical by getting them to practice drawing a diagram to show how to measure the volume of a small irregular object using a displacement (Eureka) can.

Questions involving the investigation of acceleration focus on the identification of variables (and, particularly, the control of variables) and the selection of apparatus/drawing a diagram of the apparatus. As there are two versions of this experiment, one varying resultant force and measuring acceleration, and one varying mass and measuring acceleration, you can help students to feel more confident about this practical by ensuring that they are able to describe both versions. Students should also be aware of how to describe the use of light-gates in this experiment.

Questions about investigating the behaviour of springs usually focus on measurement techniques, particularly involving ways to measure the extension of a spring directly, how to avoid uncertainties such as parallax errors, and measuring resolution. As the behaviour of school springs is elastic up to the elastic limit, and springs undergo permanent deformation above the elastic limit, this is always a favourite context for testing students' concept of range. This is also a good experiment to test students' abilities to handle and plot data. You can help students to feel more confident about this practical by giving them plenty of practice of plotting force–extension graphs.

The practical on investigating waves in a ripple tank and on a string involves two experiments. Questions about the ripple tank experiment invariably involve describing a method to measure the speed of a water wave, and require students to identify the key pieces of apparatus needed. This experiment involves quite a lot of uncertainties, so you should discuss these with students. The waves on a string experiment is a good way to test students' ability to draw diagrams and describe how measurements can be taken to determine the speed of the waves on the string. You can help students to feel more confident about this practical by getting them used to using the wave equation to determine wave speed, and to identify ways to measure wavelength and frequency.

The investigating infrared radiation experiment is one of the few physics experiments that raises any real safety concerns, as it involves the use of hot water inside a Leslie cube. You can help students to feel more confident about this practical by getting them to produce simple risk assessments involving identification of the key hazards, identification of the risks, and any control measures that can be used to minimise the risks. This experiment is also a good way to test students' abilities to draw diagrams and identify key pieces of apparatus.

SOLVE

Now turn back to the original question with the student. Work through the writing frame in the workbook together. The solutions are given below.

Design an investigation that shows the effect of increasing the mass of an object hung from a spring.

What equipment would you use for this procedure? You can show this using a diagram. List the range of masses you would use. How will you measure the extension of the spring?

Provided that the range of masses is low, springs obey the equation:

force = spring constant × extension

The force (weight) of the hanging masses is given by:

weight = mass × gravitational field strength

If the equation is correct, then adding more mass should increase the extension, and doubling the mass should double the extension.

You will need:

- 2 clamps
- 2 bosses
- 1 stand
- 1 30-cm ruler
- 1 pointer
- 1 spring
- 1 mass stack (6 × 100 g)

For a diagram, refer to key ideas card P30.6.

You can measure the extension directly off the ruler if you ensure that the zero of the ruler is in line with the pointer at the bottom of the unloaded spring. As masses are added, the spring stretches and the extension is measured off the ruler in line with the pointer.

List the steps of the practical procedure.

- Set up the apparatus as shown in the diagram.
- Add one mass to the mass hanger (100 g). Record the mass.
- Measure and record the extension of the spring using the pointer.
- Remove the mass stack from the spring. Check that the end of the spring has returned to zero on the ruler to ensure that no permanent stretching has occurred (and that the ruler has not moved).
- Repeat these steps for masses of 200 g to 600 g in 100 g steps.
- Remove the mass stack from the spring.
- Repeat these steps twice more so that you have three repeats for each added mass.
- Calculate the mean extension for each added mass.
- Plot a graph of extension (y-axis) against mass (x-axis).
- Draw a line of best fit through your points and state the relationship between the mass and the extension.

Design an investigation that shows the effect of increasing the mass of an object hung from a spring.

What are the hazards? What precautions would you take to avoid harm?

The main hazards in this experiment are:

- The mass stack falling off the spring and falling on a foot, causing a bruise or a break. The risk from this can be minimised by performing the experiment in the middle of a laboratory bench.
- The stand falling over and hitting someone in the face, causing a cut or a bruise or damaging an eye. The risk from this can be minimised by clamping the stand to the bench and wearing safety goggles.

What would you expect to happen?

If the spring obeys the equation:

force = spring constant × extension

the graph of extension (*y*-axis) against mass (*x*-axis) should be a straight line through (0, 0). Doubling the mass on the spring will double the measured extension.

Why would you expect this?

New, unstretched springs will obey the equation:

force = spring constant × extension

provided that the range of 0–600 g is not exceeded.

PRACTISE

Use the additional practice questions on this powerful idea at the back of the workbook in future sessions. Additional practice questions for Session 30 can be found in the Draw, Estimate, Justify and Sketch sections.

SESSION 1 BIOLOGICAL STRUCTURES AND THEIR FUNCTIONS

Muscle cells and nerve cells are both involved in movement. Compare the structure and function of these cells.

Which structures do muscle cells and nerve cells have in common?

- Both cells have a cell membrane, nuclei and cytoplasm.
- Both cells have lots of mitochondria.

> *Muscle cells and nerve cells are types of animal cells. You need to know that most animal cells have a nucleus, cytoplasm, a cell membrane, mitochondria, and ribosomes.*

How is the shape of each cell different?

- Nerve cells are more branched at the ends than muscle cells.
- Nerve cells are much longer than muscle cells.

> *A diagram of each cell type is shown in the workbook.*

Which structures are different in muscle cells and nerve cells?

- Muscle cells have protein filaments.
- Nerve cells have myelin.
- Nerve cells have many connections at both ends.

> *Only nerve cells contain myelin. It helps them to carry out their job.*

For each of the differences, explain how that structure or shape helps the cell do its job.

- Muscle cells have protein filaments that slide over each other to move the muscle.
- Nerve cells are very long so they can transmit impulses over a long distance.
- Nerve cells have a myelin sheath which acts as an insulator, and so makes the transmission of impulses faster.

> *The long, thin part of a nerve cell has a myelin sheath that contains gaps. Impulses can travel from one gap to the next, which allows them to travel faster.*

Now bring together your answers: What do both cells have that they use to make movement happen? What does each cell have that is different that helps it to do its specific job?

- Both cells have mitochondria that release energy – the muscle cell uses the energy to move filaments over each other, and the nerve cell uses the energy to create an electrical impulse.
- Muscle cells are specialised because they have filaments that allow them to shorten, and nerve cells are very long and have a myelin outer layer so they can conduct electrical impulses over a distance quickly.

This part of the question allows you to bring together the different parts discussed. Remember that mitochondria are used by cells to release energy. Cells that need lots of energy often have many mitochondria.

SESSION 9 REPRODUCTION AND INHERITANCE

The organisms alive today are very different from the organisms that lived millions of years ago. Peppered moths are either white and speckled, or black. During the daytime peppered moths rest on branches of trees. In the 19th century the bark of these trees became covered in black soot.

Show how peppered moths got darker in colour through natural selection when the trees they lived on became sooty.

In the original population of moths, most moths were light coloured, and a few were dark. Describe why some organisms look different.

Variation in a population is caused by mutation, meiosis and sexual reproduction. Variation is why some organisms look different.

> *There are some important keywords here: variation, mutation, meiosis. Make sure you know what they mean. Check the glossary in the workbook if you're not sure.*

When the trees got sooty, more of the dark-coloured moths survived. Why?

Dark-coloured moths were more likely to survive because it was easier for them to hide from predators on the dark sooty trees.

> *The darker moth was better adapted to the new environment than the lighter moth.*

Camouflage like this is an adaptation to not being eaten. Animals and plants have many adaptations that help them survive. Write or draw four examples below.

There are many different adaptations. Some examples include:

- Polar bears have thick fur to insulate them from the cold.
- Woodpeckers have long tongues to take insects from crevices in trees.
- Cheetahs run fast to catch prey.
- Tigers have stripes to camouflage from prey.
- Flowers have bright colours to attract pollinators.

> *The question asks for adaptations of animals and plants, so you shouldn't give examples of other living things, such as bacteria or fungi.*

More of the next generation of moths were darker coloured. Why?

More dark moths were able to avoid being eaten by predators, so they were able to reproduce and pass on the dark colouring to more offspring.

> *The moth's colour is an inherited characteristic. It is passed on from parents to their offspring. If more dark moths reproduce then there will be more dark offspring.*

Over time the whole population became darker coloured. Bring together the steps that led to this.

- Environmental change – trees became darker because of sooty pollution.
- Moths were a variety of different colours because of mutation, meiosis and sexual reproduction.
- Moths that were darker were better able to hide from predators.
- The darker moths survived and reproduced, having offspring which were also dark.
- Over the generations the population of moths grew darker.

> *This part of the question allows you to bring together the different parts discussed. Together, they describe the process of natural selection.*

SESSION 13 ATOMIC STRUCTURE

The trends in physical and chemical properties in the periodic table depend on atomic structure and electronic structure.

Compare the reactivity of magnesium and calcium.

An understanding of how the periodic table works is key to answering this question.

Find magnesium and calcium in the periodic table. Which group are they in? What does this mean?

They are both in Group 2. This means they both have two electrons in their outer shell.

Groups are columns in the periodic table. They are numbered from left to right. The group number matches the number of electrons in the outer shell. An exception to this is group 0, on the right of the periodic table, where the outer electron shell is full.

How many protons, neutrons and electrons does an atom of magnesium have? How many protons, neutrons and electrons does an atom of calcium have? Use the periodic table to find out. Write your answers in the table.

	No. of protons	No. of neutrons	No. of electrons
Magnesium	12	12	12
Calcium	20	20	20

You need to know how to calculate the number of neutrons from the mass number and atomic number. It is usually different from the number of protons and electrons. You will have a periodic table in the exam.

Draw the electronic structure of calcium and magnesium. Use crosses for electrons. Label the energy levels.

Magnesium:

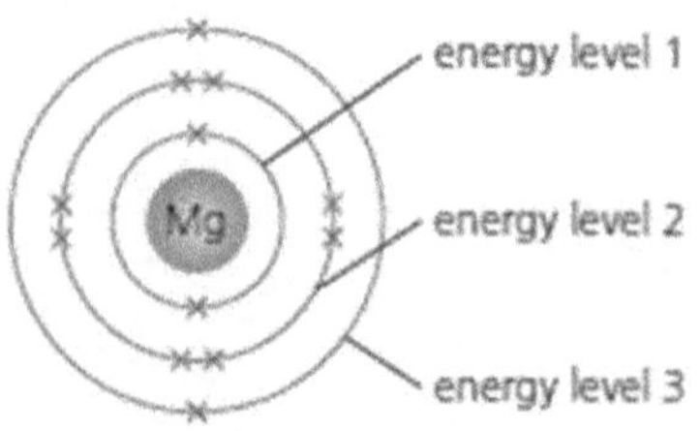

Calcium:

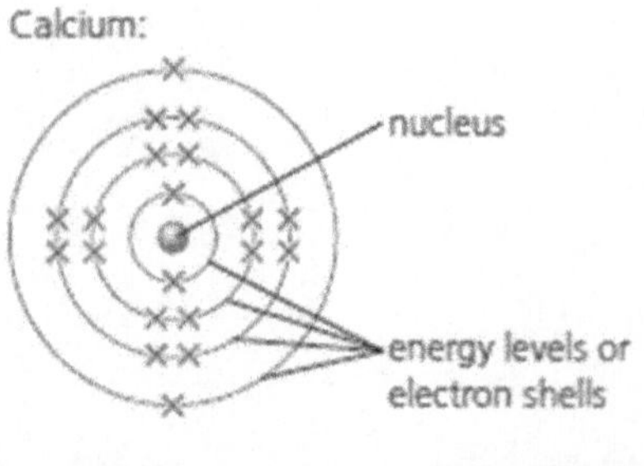

The first energy level is closest to the nucleus.

Magnesium is above calcium in the periodic table. Which one is more reactive? Why?

Calcium is more reactive. Down the group, the outer electrons are further away from the nucleus. Therefore, the electrostatic attraction with the nucleus is weaker and so it is easier to transfer electrons during a chemical reaction.

Try to use relevant key terms in your answer. Electrostatic attraction is the force of attraction between opposite charges.

Bring together your previous answers to make a list of how magnesium and calcium are similar.

- Both are metals.
- Both are in Group 2 of the periodic table.
- Both have 2 electrons in their outer shell.

Bring together your previous answers to make a list of how magnesium and calcium are different.

- Calcium has more electron shells or energy levels.
- In calcium, the outer shell electrons are further away from the nucleus, so the electrostatic attraction with the nucleus is weaker.
- Calcium is therefore more reactive than magnesium.

This part of the question allows you to bring together the different parts discussed.

SESSION 19 ELECTROLYSIS

a) Describe how pure aluminium is extracted from aluminium ore.

b) Explain why aluminium is so expensive.

For part a): Where is aluminium found naturally?

Aluminium is found in the ground as aluminium ore or aluminium oxide.

> *Remember that metal ores must be mined before the metal can be extracted.*

For part a): Why can aluminium not be extracted using carbon like iron can?

Aluminium is too reactive.

> *Aluminium is above carbon in the reactivity series. Only metals that are less reactive than carbon can be extracted using carbon.*

For part a): What are the names of the two electrodes used in electrolysis?

Positive electrode: anode. Negative electrode: cathode.

> *Try not to forget these names or get them confused. You could use a mnemonic to remember which is which.*

For part a): Add labels to the diagram to show what happens at each electrode in electrolysis.

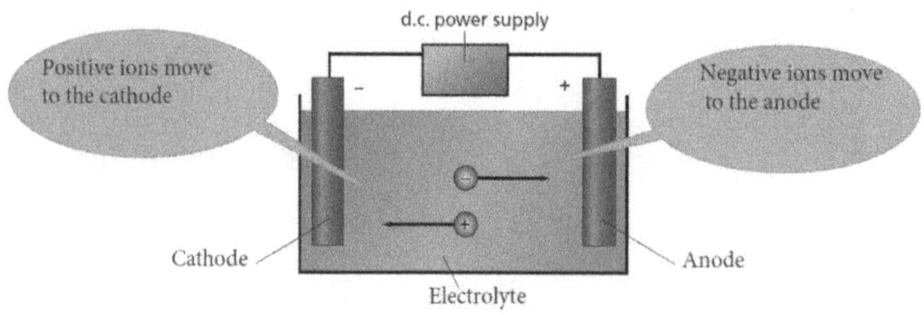

For part a): Describe or draw how electrolysis is used to extract aluminium. Add as many labels as you can.

- Aluminium oxide is mixed with cryolite and heated until it melts.
- Cryolite lowers the melting point.

> *Remember that cryolite does not take part in the reaction. It acts as an impurity to lower the melting point of aluminium oxide.*

- An electric current is then passed through the liquid.
- Al^{3+} ions are attracted towards the cathode where the molten aluminium forms, which is then tapped off.
- The oxide ions are attracted to the anode where oxygen gas is formed.

OR Labelled diagram

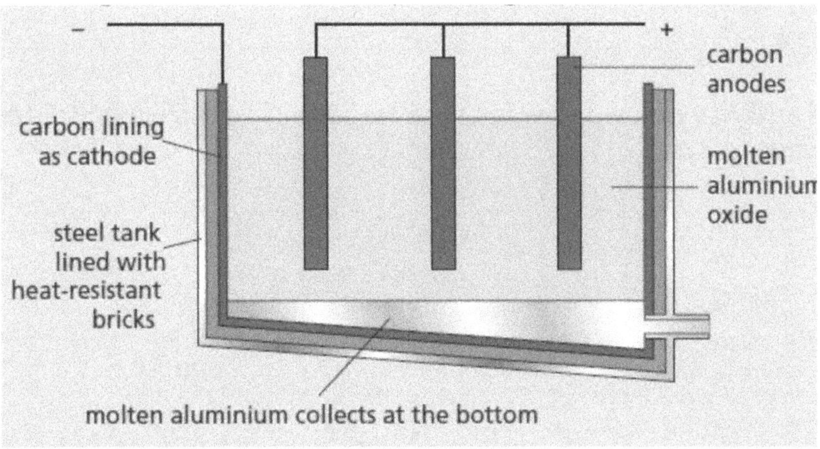

You cannot use a general electrolysis diagram here. It must show the industrial process in the question.

For part b): Give two reasons why a lot of energy is used during the extraction of aluminium.

- To melt the aluminium oxide.
- Electricity is need for electrolysis itself.

Large amounts of energy are needed for all metal extraction processes that use electrolysis.

For part b): Why do the anodes need to be replaced regularly?

As soon as the oxygen is formed it reacts with the carbon to form carbon dioxide, removing the carbon from the anode.

Remembering that oxygen forms at the anodes is key to answering this part of the question.

For part b): Now bring it together: Why is aluminium so expensive?

It takes a lot of energy to produce aluminium. First, to melt the aluminium oxide, and then for the electrolysis itself. The carbon anodes regularly need to be replaced because as soon as the oxygen is formed it reacts with the carbon to form carbon dioxide; so the carbon anodes need to be replaced, which adds to the expense.

For a good answer, you should bring together all the main ideas discussed.

SESSION 25 WAVES

The diagram shows a transverse water wave. The frequency of the wave is 3 Hz and the wavelength of the wave is 2 m.

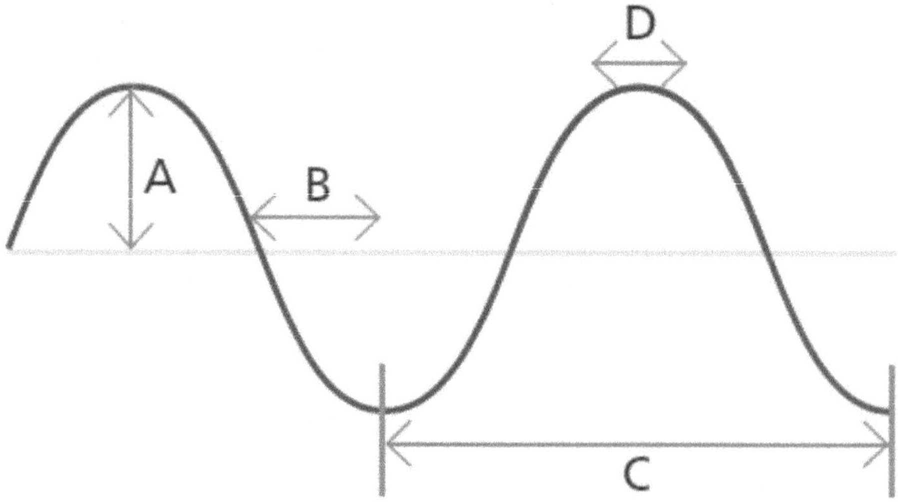

a. **Identify the label corresponding to the:**

  i. **amplitude**

  ii. **wavelength.**

b. **Calculate the speed of the wave using the wave equation.**

c. **The table below shows the parts of the electromagnetic spectrum. Some of the parts are missing.**

radio waves		**infrared**		**ultraviolet**		**gamma rays**

  i. **Write the missing words in the table.**

  ii. **Draw an arrow under the table to show in which direction wavelength increases.**

  iii. **Explain why ultraviolet rays are dangerous to humans.**

For part a): What are the characteristics of a transverse (e.g. water) wave? Where do we measure amplitude and wavelength from?

  i. Amplitude = A

> *This is the only letter label that shows displacement away from the undisturbed position (the yellow line).*

  ii. Wavelength = C

> *This is the only letter label that shows a feature on one wave and the next wave.*

For part b): Gather all the numerical information from the question that you need to use the wave equation:

wave speed (m/s) = frequency (Hz) × wavelength (m)

Then substitute the numbers into the equation and use a calculator to work out the answer. Repeat this on your calculator to check that you have entered the numbers correctly. What are the units of wave speed in this question?

wave speed = frequency × wavelength

> *This is one of the equations that you need to learn – you could be asked to recall it during the exam.*

= 3 Hz × 2 m

> *Always write out your working – it is then easier to spot any silly mistakes.*

= 6 m/s

> *Always check your answer twice using your calculator. Try to remember units. Speeds are always given in m/s in the exam.*

For part c) i), ii): List all of the parts of the electromagnetic spectrum in the correct order. Try to remember which have the longest wavelength, and so, which way the arrow should point.

i.

radio waves	microwaves	**infrared**	**visible light**	ultraviolet	**X-rays**	gamma rays

> *Try to remember these in order.*

ii. A horizontal arrow pointing to the left should be drawn under the table.

> *Increasing frequency means decreasing wavelength, and vice versa.*

For part c) iii): You need to ***explain*** *in this question – this means you need to state the reasons for something. First, state the property of ultraviolet rays that makes them dangerous. Next, give the reasons why this makes them dangerous.*

Ultraviolet rays are dangerous to humans because they have a high energy and can harm or kill cells.

> *Ultraviolet rays are one of the three ionising forms of radiation, which means they are hazardous to living cells.*

SESSION 29 ATOMS AND RADIOACTIVITY

The table below shows how the activity of a radioactive sample of lead-217 (atomic number 82) changes over time.

Time in s	**0**	**10**	**20**	**30**	**40**	**50**
Activity in Bq	**120**	**85**	**60**	**42**	**30**	**21**

a. **Plot a graph of activity against time.**

b. **Draw a line of best fit through the points.**

c. **Use information from the graph to determine the half-life of lead-217.**

d. **Lead-217 decays by beta particle emission into an isotope of bismuth, Bi. Balance the nuclear decay equation for this decay:**

$$^{217}_{82}\mathrm{Pb} \rightarrow {}^{0}_{-1}\mathrm{e} + {}^{\cdots}_{\cdots}\mathrm{Bi}$$

For part a): You will first need to determine the axes for the decay graph. Which quantity goes on which axis? What scale do you need – what are the lowest and the highest values that you need to plot on each axis? Do you need to start the graph axes at (0, 0)? Make sure that you plot the points accurately using a sharp pencil. *Double-check them.*

For part b): Think about the general shape plotted out by the points (is it a curve or a straight line?). Remember, a line of best fit does not need to go through all the points, but it does need to be a smooth line, not join-the-dots.

a) and b)

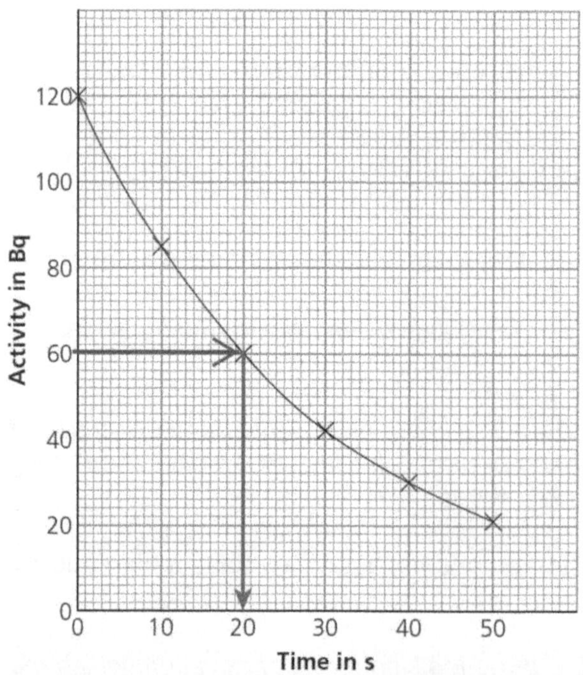

a) The axes on the graph are: Activity in Bq (on the y-axis), Time in s (on the x-axis). The graph starts at (0, 0). The y-axis goes from 0 to 120 Bq. The x-axis goes from 0 to 50 s.

b) In this case, if plotted correctly, all the points fall on a smooth curve.

For part c): The half-life is the time taken for the initial activity to fall to half its value. What is the initial activity? What is half of this initial value? Draw a straight line across the graph from this value, so it touches the line of best fit. Then draw a line down to the time axis and read off the half-life. What are the units?

The horizontal arrow shows half the initial activity, this hits the line of best fit at a time of 20 s. This is the half-life.

> *Start at the initial value, 120 Bq. Half of this is 60 Bq. Draw a horizontal line across from 60 Bq, until it touches the curve, then draw another vertical line down from this point. Where the vertical line crosses the time axis, this is the half-life.*

For part d): You need to balance the nuclear decay equation. The top numbers, mass numbers, need to be the same on each side of the equation. What is the mass number of the bismuth isotope? The bottom numbers, atomic numbers, also need to balance on either side of the equation. Remember the value for a beta particle is −1, so the atomic number of bismuth – 1 = atomic number of lead. What is the atomic number of the bismuth isotope?

$$^{217}_{82}\mathrm{Pb} \rightarrow {}^{0}_{-1}\mathrm{e} + {}^{217}_{83}\mathrm{Bi}$$

The missing mass number is 217; the missing atomic number is 83.

> *During beta decay, the mass number stays the same, but the atomic number goes up by 1. Using this rule, the top number must remain as 217, and the bottom number must go up by 1 to 83.*

Session 1: Biological structures and their functions

Collins
Secure Science

Biological structures and their functions

B1.1

Session 1: Biological structures and their functions

The structure of cells and **tissues** in living things allows them to do particular jobs to keep the organism alive. **Th**e shape and parts of cells help them carry out their **function**. Many exam questions need you to make a link between the function of a cell and its structure. For example, a plant **palisade** (leaf) **cell** has lots of chloroplasts so it can use energy from the Sun in **photosynthesis**.

Specialised human cells

Collins
Secure Science

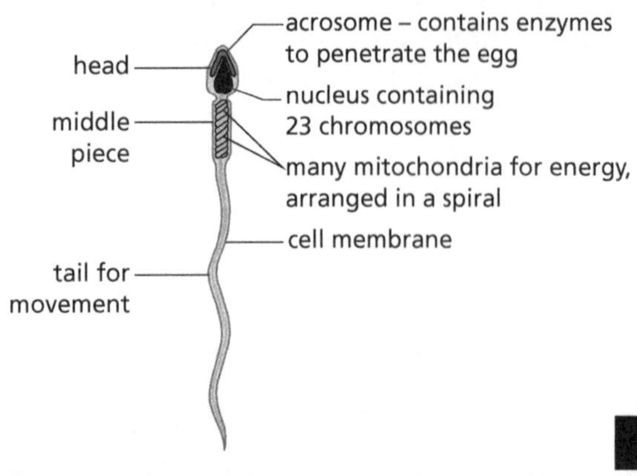

B1.2

Specialised human cells

The function of a sperm cell is to swim through the female reproductive system to fertilise an egg. Sperm cells have many **mitochondria** for energy for swimming, a tail to move and only half the genetic material of a body cell so that the fertilised egg has the right amount of genetic material.

Specialised human cells

Collins
Secure Science

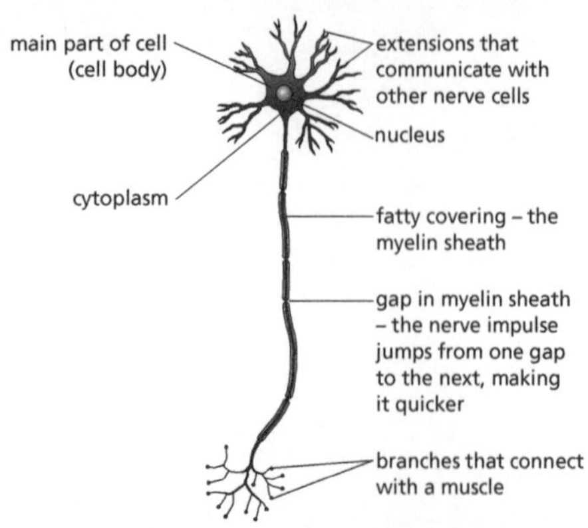

B1.3

Specialised human cells

The function of a nerve cell is to carry electrical impulses around the body. It is very long, so it can carry impulses over a distance. It has an insulating layer of **myelin** that makes conducting (carrying) impulses faster. It has lots of connections at each end so it can receive and pass on impulses.

Specialised human cells

Collins
Secure Science

B1.4

Specialised human cells

The function of a muscle cell is to shorten and lengthen so the organism can move. The muscle cell has lots of mitochondria to release energy, and has filaments that slide over each other to shorten and lengthen. The filaments make the cells look stripy.

Plant and animal cells

Collins
Secure Science

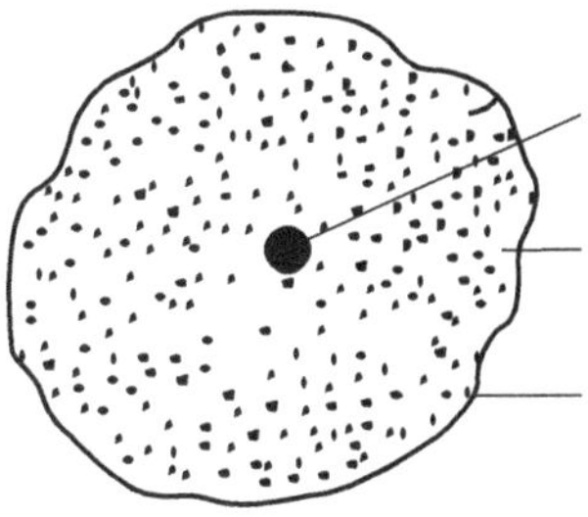

B1.5

Plant and animal cells

Collins
Secure Science

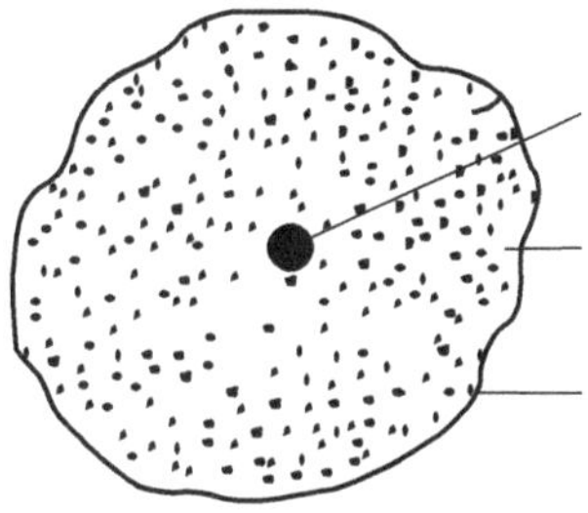

Plant and animal cells both have:

nucleus: controls the activities of the cell

cytoplasm: where chemical reactions happen

cell membrane: controls the movement of molecules into and out of the cell

B1.5

Plant and animal cells

Collins
Secure Science

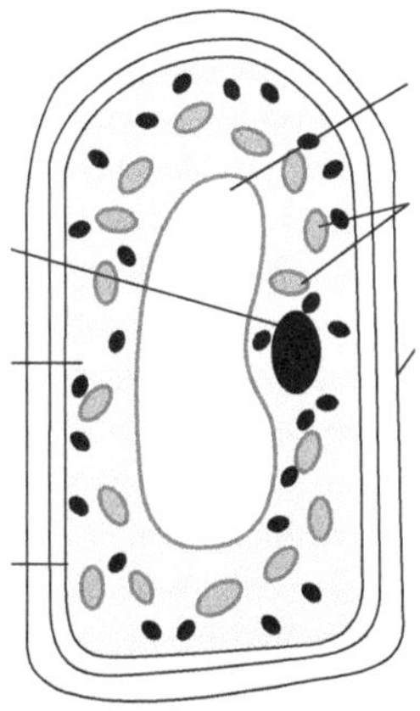

B1.6

Plant and animal cells

Collins
Secure Science

Plant and animal cells both have:

nucleus: controls the activities of the cell

cytoplasm: where chemical reactions happen

cell membrane: controls the movement of molecules into and out of the cell

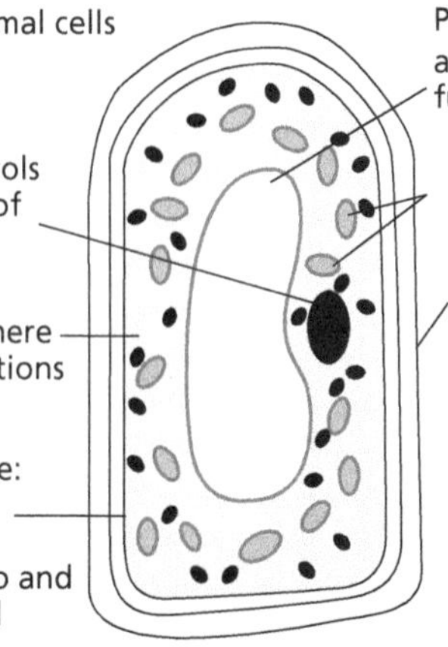

Plant cells also have:

a permanent vacuole full of cell sap

chloroplasts containing chlorophyll that absorbs light for photosynthesis

cell wall: a layer made from cellulose outside of the cell membrane that strengthens the cell

When the cell has lots of water the vacuole pushes against the cell wall to give the plant structure.

B1.6

Cell structures

Collins
Secure Science

B1.7

Cell structures

Collins
Secure Science

The structures found in cells give us clues about their functions. Cells with chloroplasts carry out photosynthesis. Cells with lots of mitochondria carry out high-energy functions such as moving. Cells with **ribosomes** make proteins such as **enzymes** used in digestion.

These structures are very small. Chloroplasts are the largest and can be seen under a light microscope, but mitochondria and ribosomes can only be seen using an electron microscope.

B1.7

Specialised plant cells

Collins
Secure Science

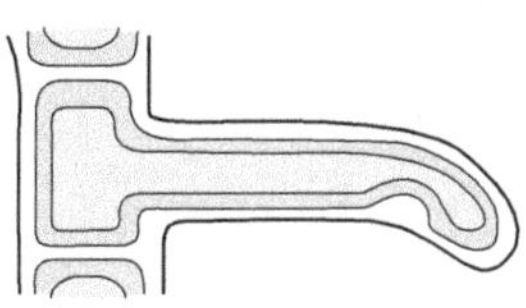

root hair cell

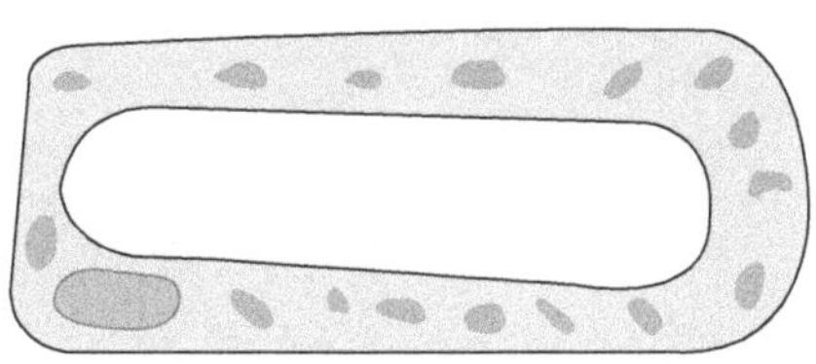

leaf cell

B1.8

Specialised plant cells

Collins
Secure Science

Root hair cells absorb water and minerals from the soil. They have a root hair which has a large area of cell membrane so lots of water can move into the cell. They have lots of mitochondria to provide energy for **active transport**. Energy is needed to push minerals into the cell when there is more of the mineral inside the plant than there is in the soil outside of the plant.

The function of a leaf cell is to photosynthesise to make glucose for the plant. Leaf cells have lots of **chloroplasts** full of **chlorophyll**. Chlorophyll absorbs light. Xylem and Phloem are specialised for transport. Xylem are hollow cells strengthened by lignin which transport water and mineral ions. Phloem are elongated cells which transport dissolved sugars from the leaves to the rest of the plant.

B1.8

Session 2: Moving molecules in living things

Collins
Secure Science

Moving molecules in living things

B2.1

Session 2: Moving molecules in living things

Coll
Secure S

Useful substances need to move through multicellular organisms. Molecules move from areas of high concentration to areas of low concentration, along a concentration gradient. When water moves across cell membranes it is called osmosis. Molecules move by **diffusion** faster when: they move over a short distance with a large surface area, there is a big difference in concentration (a high **concentration gradient**), and there is a good blood supply (in animals only).

B2

Diffusion

Collins
Secure Science

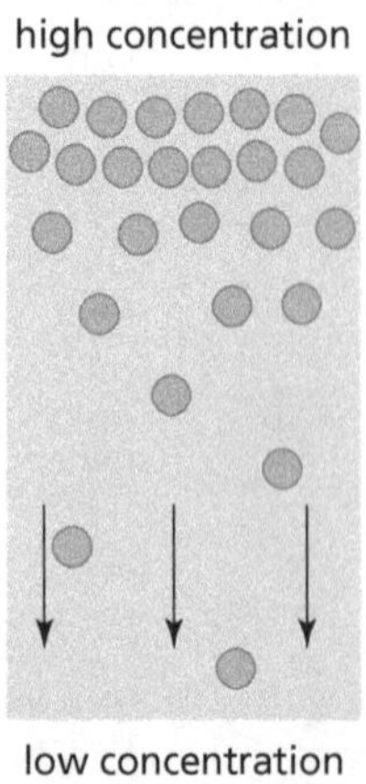

B2.2

Diffusion

Coll
Secure S

Diffusion is the movement of particles from an area of high concentration to an area of low concentration along a concentration gradient.

B2

Osmosis

Collins
Secure Science

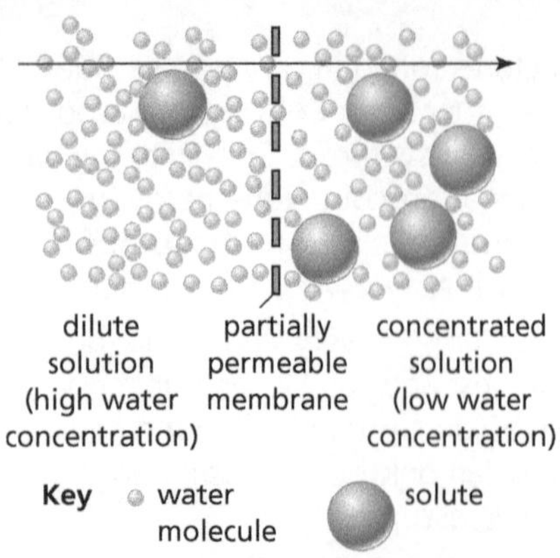

B2.3

Osmosis

Coll
Secure S

Osmosis is a special case of diffusion that happens in living things. A dilute solution contains many water molecules (a high concentration of water molecules). A concentrated solution contains fewer water molecules. Osmosis is the movement of water molecules from a dilute to a concentrated solution, across a **partially permeable membrane**.

B2

Diffusion in plants

Collins
Secure Science

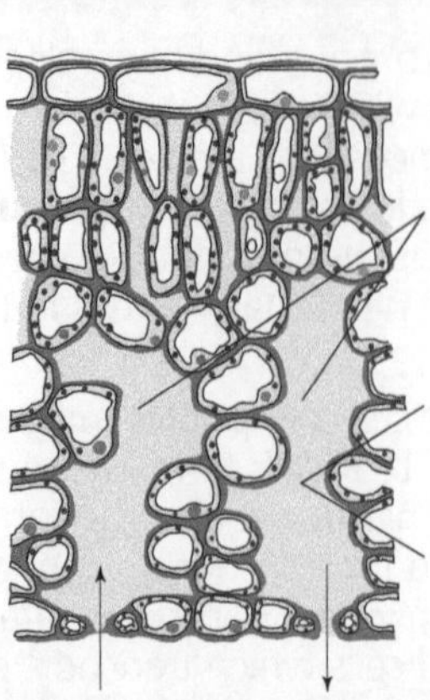

B2.4

Diffusion in plants

Coll
Secure S

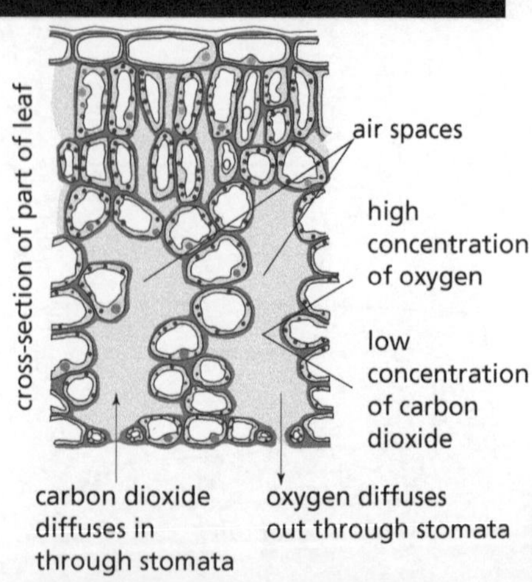

B2

Osmosis in animal cells

Collins
Secure Science

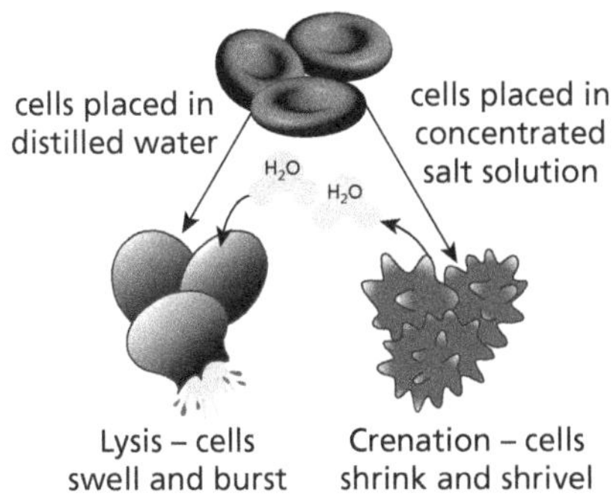

B2.5

Osmosis in animal cells

Collins
Secure Science

Animal cells do not have a cell wall. Therefore, when too much water moves into animal cells by osmosis they will burst. This is called lysis. If too much water moves out of animal cells by osmosis they will shrink and shrivel up.

B2.5

The structure of plants

Collins
Secure Science

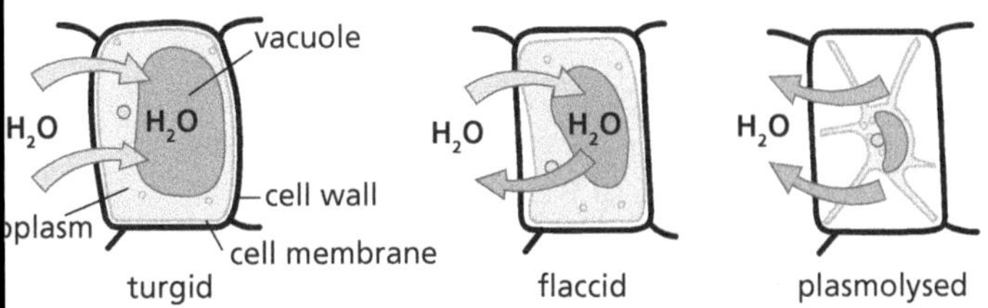

B2.6

The structure of plants

Collins
Secure Science

The cell wall and the vacuole work together to give the plant structure. When there is enough water available the vacuole pushes against the cell wall. The cell is **turgid**. When there is not enough water in the vacuole it shrinks and pulls away from the cell wall. The cell is **flaccid**. When there is not enough water in the plant it **wilts**.

B2.6

Moving water through a plant

Collins
Secure Science

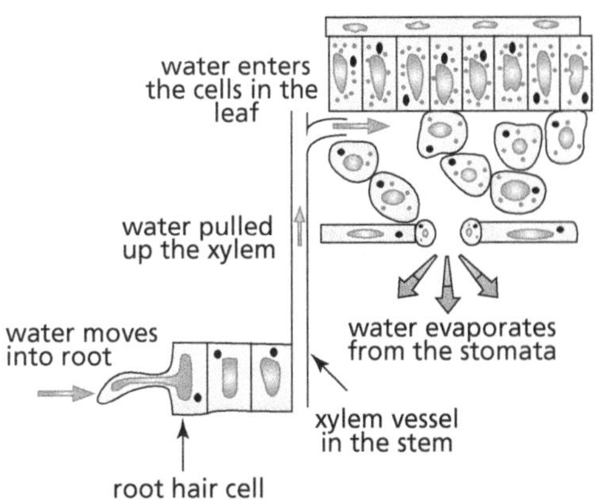

B2.7

Moving water through a plant

Collins
Secure Science

Water moves through a plant in different ways. The leaves use up water in photosynthesis, so there is a low concentration of water inside the plant compared with the soil. Water moves into the root by osmosis and up through the stem towards the leaves in a process called **transpiration**. The water is pulled because water is leaving the leaf by evaporation.

B2.7

Session 3: Life systems

Collins
Secure Science

Life systems

B3.1

Session 3: Life systems

Collins
Secure Science

Useful substances need to move through multicellular organisms. Molecules move by diffusion, osmosis or active transport. Molecules move across **specialised exchange surfaces** such as the small intestine. Exchange surfaces are adapted to allow molecules to be transported quickly. An **efficient** exchange surface has a large surface area, a thin membrane, a good blood supply (in animals only) and is well **ventilated** (for gas exchange in animals).

B3.1

Lung and gas exchange

Collins Secure Science

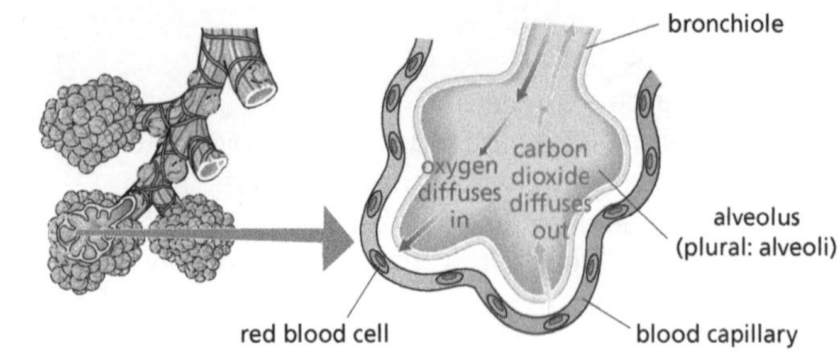

B3.2

Lung and gas exchange

Coll
Secure S

The lungs and surrounding tissues are adapted for **gas exchange**. Oxygen needed for cell respiration diffuses from the air into the blood. Carbon dioxide made by cell respiration diffuses from the blood into the air. The many small spheres of **alveoli** and the dense network of capillaries increase the surface area for diffusion. The walls of the alveoli and the capillary are one cell thick giving a short distance. The surface of the alveoli is moist allowing gases to dissolve and diffuse more easily. Blood movement and breathing maintain the concentration gradients needed for diffusion.

B3

The digestive system

Collins Secure Science

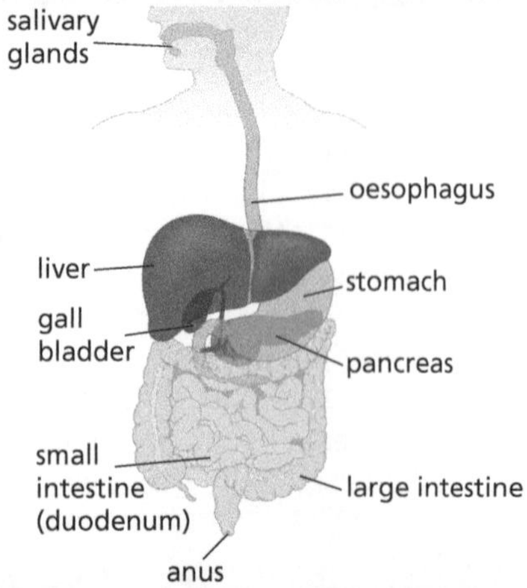

B3.3

The digestive system

Coll
Secure S

The digestive system is a group of organs that break down food into smaller soluble products which can be absorbed into the blood. Many of the organs release enzymes which act as biological catalysts, speeding up the rate of digestion.

B3

Enzyme action

Collins Secure Science

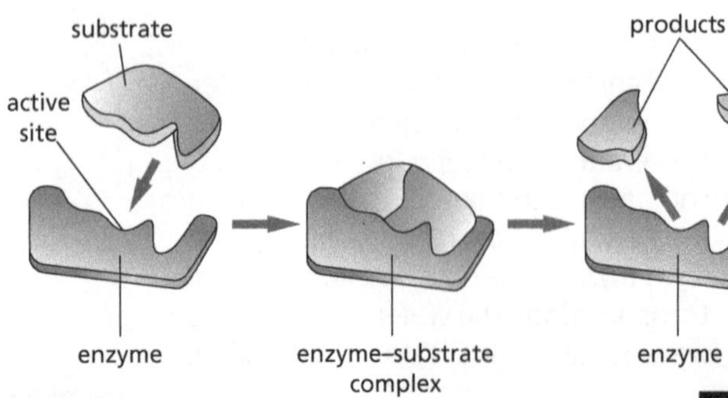

B3.4

Enzyme action

Coll
Secure S

Enzymes are biological **catalysts** that speed up reactions. The **lock and key theory** is a simple model to explain how enzymes work. The **substrate** is the molecule that 'fits' into the **active site** of the enzyme like a key 'fits' into a lock. When the substrate binds to the active site an enzyme–substrate complex is formed. The enzyme then changes the substrate into the products.

B3

Enzyme and digestion

Collins Secure Science

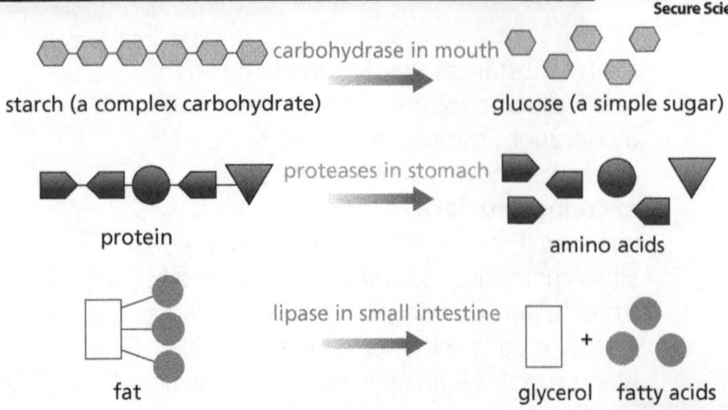

B3.5

Enzyme and digestion

Coll
Secure S

The digestive system is a connected system with the function of breaking down food for absorption. **Digestive enzymes** are released into the digestive system and work outside of cells to digest food into small soluble molecules. These small molecules are then absorbed into the blood in the small intestine. Carbohydrate digestion starts in the mouth with **amylase** (a type of carbohydrase) and then continues in the small intestine. Protein digestion starts in the stomach with **proteases** and then continues in the small intestine. Fat digestion starts in the small intestine with **lipase**.

B3

Absorption in the small intestine

Collins
Secure Science

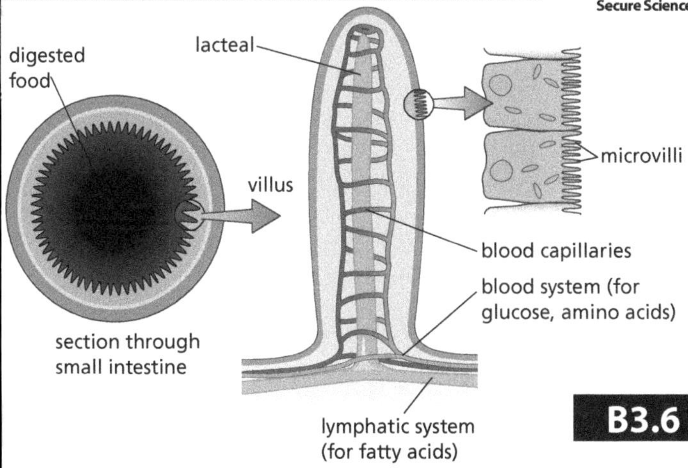

B3.6

Absorption in the small intestine

Collins
Secure Science

Absorption is the process by which soluble products of digestion move into the blood from the small intestine. The small intestine is about 7 m long, giving lots of time for absorption as food travels along. The inside surface (lumen) of the small intestine has **villi** which are made of cells covered in **microvilli**, which increase the surface area for absorption. Blood capillaries in the villi transport molecules away in the blood keeping the concentration gradients needed for absorption.

B3.6

Staying alive

Collins
Secure Science

Staying alive

B4.1

Staying alive

Collins
Secure Science

Cells are the building blocks of living organisms. Cells with similar structure and function are grouped together into **tissues**. Tissues with a similar function work together to form **organs**. Organs are grouped into organ systems which have an overall function. For example, they move substances around the body. Organ systems work together to keep living organisms alive. Damage to cells, tissues or organs within body organ systems can cause disease or death.

B4.1

The respiratory system

Collins
Secure Science

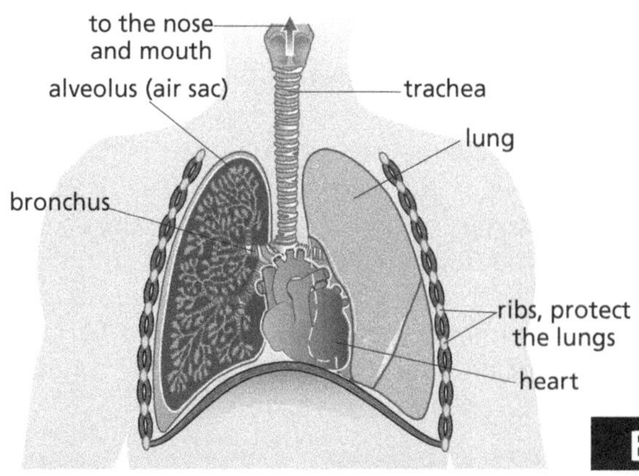

B4.2

The respiratory system

Collins
Secure Science

The respiratory system is made of organs and tissues that work together to get air into and out of the body. In the lungs oxygen diffuses into the body and carbon dioxide diffuses out of the body. The **trachea** and **bronchi** are tubes through which air passes into and out of the lungs. This allows gas exchange to happen in the alveoli, which are small sacs with thin walls that are surrounded by a **capillary network**.

B4.2

Structure of the heart

Collins
Secure Science

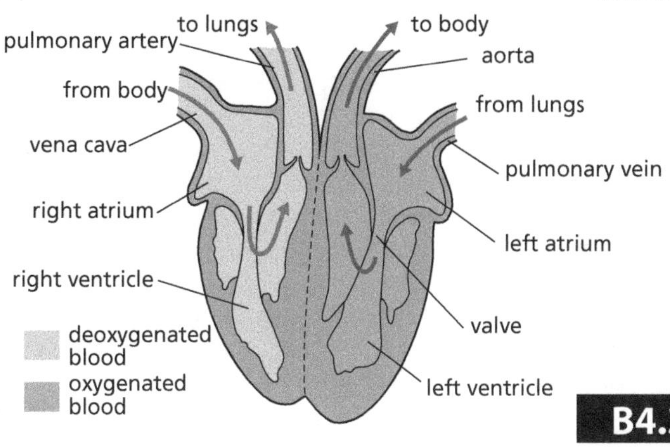

B4.3

Structure of the heart

Collins
Secure Science

The heart is an organ made of specialised **cardiac muscle** tissue. The right side of the heart receives blood from the body through the **vena cava** into the right **atrium**. Then, the right **ventricle** pumps blood to the lungs through the **pulmonary artery**. In the lungs, gas exchange happens. The oxygenated blood then returns to the left atrium through the **pulmonary vein**. The left ventricle then pumps the blood to the body through the **aorta**, this has the thickest wall because it pumps blood a greater distance around the rest of the body. The heart can increase and decrease the heart rate depending on what the body needs. When we exercise more oxygen is needed so the heart pumps the blood faster.

B4.3

Blood and blood vessels

Collins
Secure Science

Parts of blood

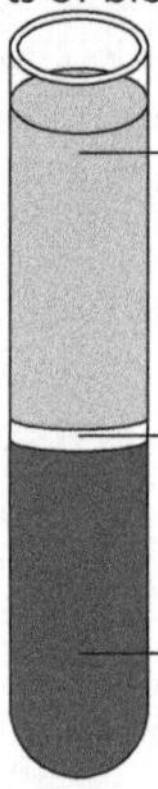

B4.4

Blood and blood vessels

Coll
Secure S

The circulatory system carries blood around the body. Blood is carried in three main types of vessel – arteries, veins and capillaries. These vessels are specialised to their function, as shown in the table. **Coronary arteries** supply the heart muscle with blood. **Coronary heart disease** is caused by damage to the coronary arteries or by fatty deposits forming in the vessels.

B4.

Blood and blood vessels

Collins
Secure Science

Arteries	**Veins**	**Capillaries**
thick, elastic wall, not **permeable**; small lumen 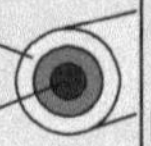	thin wall, not permeable; large lumen; valve – prevents backflow of blood 	wall is one cell thick and permeable to allow oxygen to diffuse into tissues 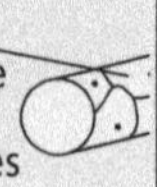
carry blood away from the heart	carry blood to the heart	carry blood from arteries to tissues and then join up with veins
carry blood under high pressure with a pulse	carry blood under low pressure, blood flows smoothly	pressure falls and pulse disappears

B4.5

The circulatory system

Collins
Secure Science

B4.6

The circulatory system

Coll
Secure S

The **circulatory system** is a transport system made of the heart and blood vessels. It is a double system. One system pumps blood from the heart to the lungs and back. This system gets oxygen into the blood and carbon dioxide out. The other system pumps blood from the heart to the body and back. This provides cells with the oxygen and nutrients they need (for respiration) as well as taking away the carbon dioxide (produced in respiration).

B4

Nervous control

Collins
Secure Science

Nervous control

B5.1

Nervous control

Coll
Secure S

Our bodies have control systems that communicate with parts of the body to make changes. These changes keep our internal environment stable so that the processes that keep us alive work properly. In nervous control, changes are detected by **receptors**. Receptors send impulses to the **coordination centre** (brain or spinal cord) via an electrical signal in nerves. The brain then coordinates an action. The action is sent in nerves to an **effector** organ. Nervous control is fast- and short-acting.

B5

The nervous system

Collins
Secure Science

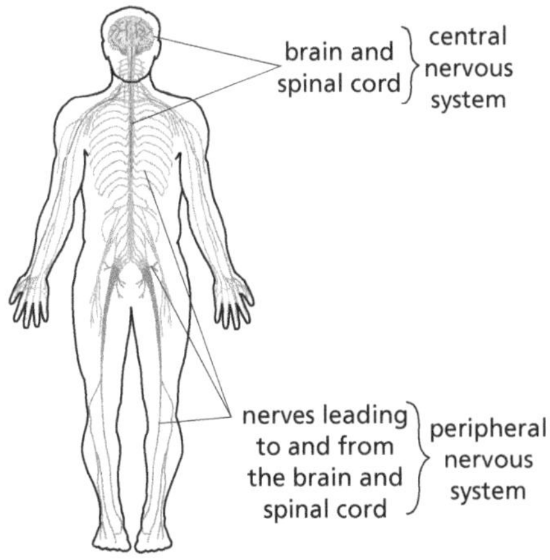

B5.2

The nervous system

Collins
Secure Science

The nervous system is made of the **central nervous system** (CNS) and the **peripheral** nervous system. The CNS is made of the brain and the spinal cord. The CNS coordinates the response to a **stimulus** – a change in the environment.

B5.2

Coordinate responses

Collins
Secure Science

RECEPTOR
…tects changes
…e environment

→ message →

COORDINATION CENTRE
e.g. the brain, spinal cord or a gland, (e.g. the pancreas) receives and processes information

→ message →

EFFECTOR
– muscles that bring about a response, or
– a gland that secretes a hormone to restore optimum levels or bring about a response

B5.3

Coordinate responses

Collins
Secure Science

Sensory neurones connect receptors to the CNS. Receptors are special areas found in our sense organs which detect a stimulus. For example, light receptors are found in our eyes; temperature receptors are found in our skin. Motor neurones connect the CNS to effectors. Effectors are muscles or glands and allow our bodies to respond to changes in the environment. For example, when we smell food the salivary glands in our mouth release saliva.

B5.3

The reflex arc

Collins
Secure Science

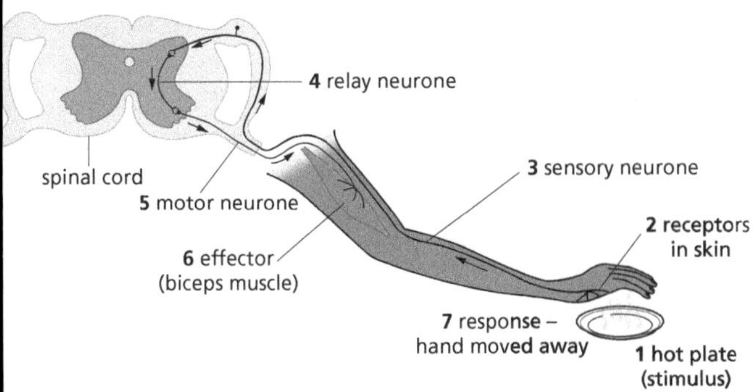

B5.4

The reflex arc

Collins
Secure Science

Reflex actions are rapid, automatic responses to a stimulus. In simpler organisms they are the basis of behaviour. In humans they stop us from getting hurt. Reflex actions include closing our eyes if an object comes near to them, and the grasping reflex of a baby gripping a finger. The diagram shows a reflex action to touching a hot plate. The stimulus (touching the plate) is detected by a receptor (in the skin). The impulse is then sent to the CNS through a sensory neurone. The **relay neurones** in the CNS bring about a response and the impulse is sent to an effector (the biceps muscle) through a motor neurone. The reflex action makes sure the person quickly moves their hand away from the hot plate. The stimulus relay neurones in the CNS (*normally the spinal cord*).

B5.4

Structure of nerve cells

Collins
Secure Science

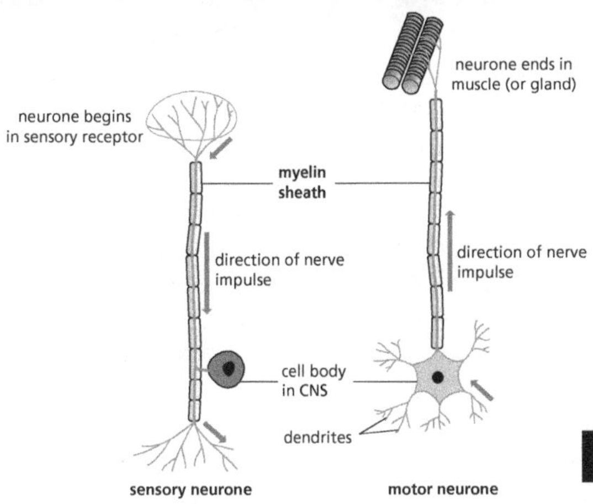

B5.5

Structure of nerve cells

Collins
Secure Science

Nerve cells (also called **neurones**) are the specialised cells of the nervous system. Nerve cells can transmit electrical impulses. At the ends of the nerve cell are fine branches called dendrites that help neurones connect to other cells. Nerve cells are also specialised to be very long and have a fatty layer of myelin. The length helps the nerve cells transmit impulses around the body, and the myelin insulates the nerve so it can transmit information quickly. **Sensory neurones** carry information about changes in the environment. **Motor neurones** carry information about how to respond to the environment.

B5.5

Hormonal control

Collins
Secure Science

Hormonal control

B6.1

Hormonal control

Coll
Secure S

Our bodies have control systems that communicate with parts of the body to make changes. These changes keep our internal environment stable so that the processes that keep us alive work properly. In hormonal control, receptors sense when a **hormone** needs to be produced. Hormones are secreted by glands and travel through the blood to the organs where they have an effect. Hormones act more slowly than nerves, but their effects last longer.

B6

The endocrine system

Collins
Secure Science

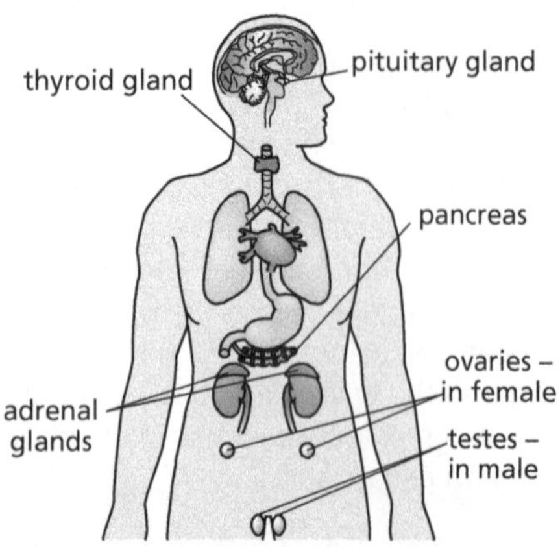

B6.2

The endocrine system

Coll
Secure

The **endocrine system** is made up of glands that make specific chemical messengers called hormones which are released into the blood.

B6

The endocrine system

Collins
Secure Science

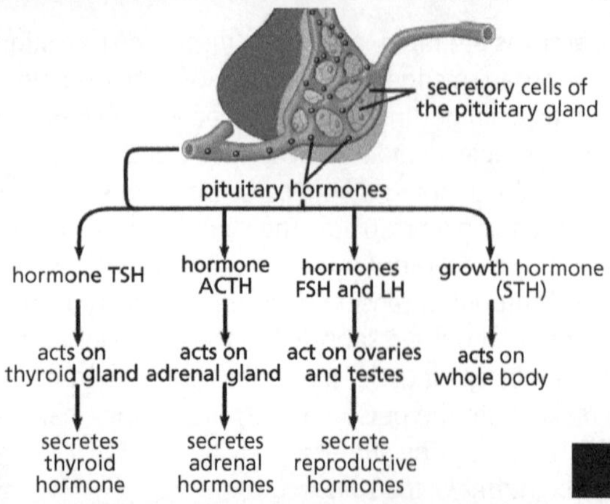

B6.3

The endocrine system

Coll
Secure

The pituitary gland sits just below the brain and is described as the 'master gland'. The pituitary secretes some hormones that have a direct effect on target organs; for example, growth hormone. The pituitary also secretes hormones that have an indirect effect by acting on other glands to secrete their hormones into the blood.

B6

Controlling blood sugar

Collins
Secure Science

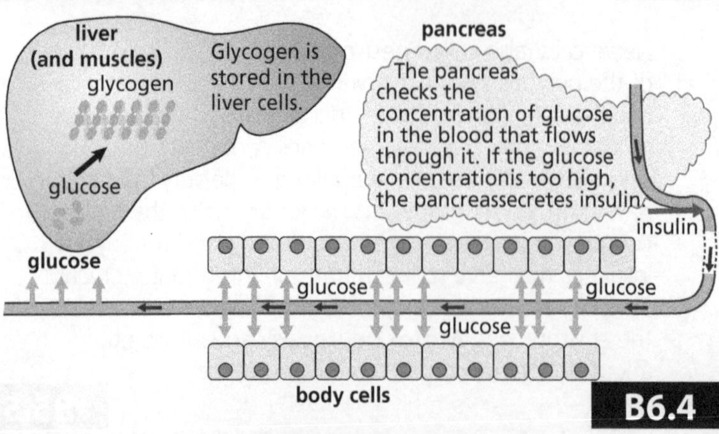

B6.4

Controlling blood sugar

Coll
Secure

Blood sugar (glucose) is controlled by the **pancreas**. The pancreas detects changes and releases the hormone insulin i sugar levels in the blood get too high. Insulin makes glucose go into cells to be used in cell respiration, or into the liver or muscles to be stored as the carbohydrate **glycogen**. People with diabetes cannot control their blood glucose levels. In **Type 1 diabetes** the pancreas cannot produce enough insulin. This causes high blood sugar levels that are treated by injecting insulin. In **Type 2 diabetes** the cells no longer respond to insulin and sometimes the body cannot make enough. Type 2 diabetes can be treated by a carbohydrate-controlled-diet, increased exercise and, if necessary, medication, including insulin. Obesity is a risk factor for Type 2 diabetes.

B6

Puberty in females

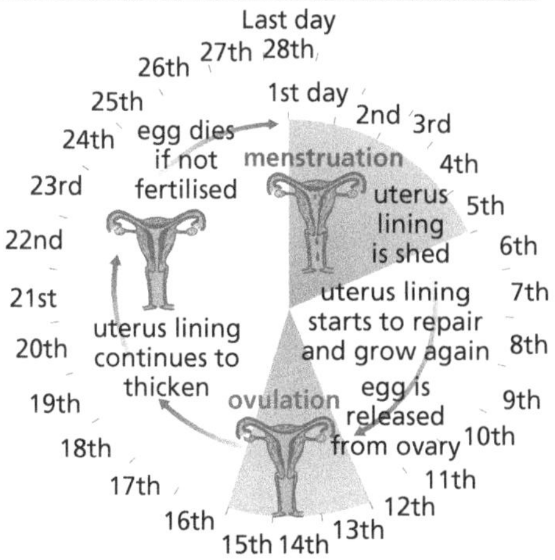

B6.5

Puberty in females

During puberty hormones are released which help the body develop **secondary sex characteristics**. In females the **menstrual cycle** begins. **Follicle stimulating hormone** (FSH) causes an egg to mature in the ovary. **Luteinising hormone** (LH) causes the ovary to release the mature egg. **Oestrogen** and **progesterone** both help to repair and maintain the lining of the **uterus**. To prevent pregnancy **contraception** can be used. Barrier methods (condom and diaphragm) prevent the sperm from reaching the egg. Hormonal methods (implant, patch, pill) stop eggs from being released by inhibiting FSH.

B6.5

Homeostasis

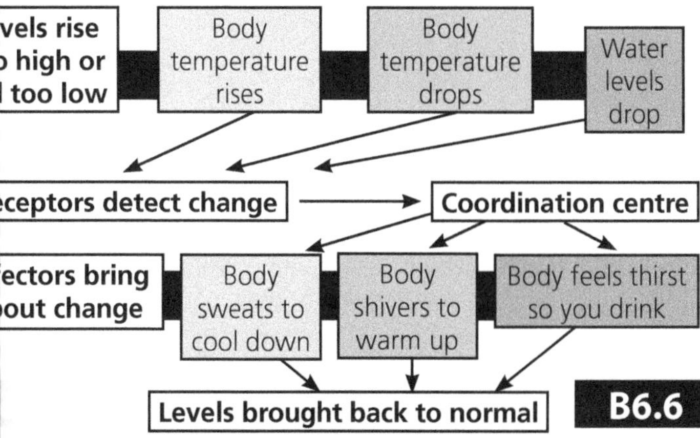

B6.6

Homeostasis

Homeostasis is the control of conditions inside the body between narrow limits. This control is automatic and is carried out by the nervous and endocrine systems. Water levels and body temperature have to be kept stable for the optimum conditions for enzymes to work.

B6.6

Disease and immunity

Disease and immunity

B7.1

Disease and immunity

Pathogens are microbes that cause disease by destroying body cells and/or producing toxins. Pathogens have different proteins from body cells. **White blood cells** recognise these proteins as **foreign**. The body makes white blood cells that can attach to the foreign proteins and destroy them. Some of the matching white blood cells remain in the body after the pathogen is destroyed. If the pathogen enters the body again, the white blood cells can multiply quickly and prevent disease.

B7.1

The immune system

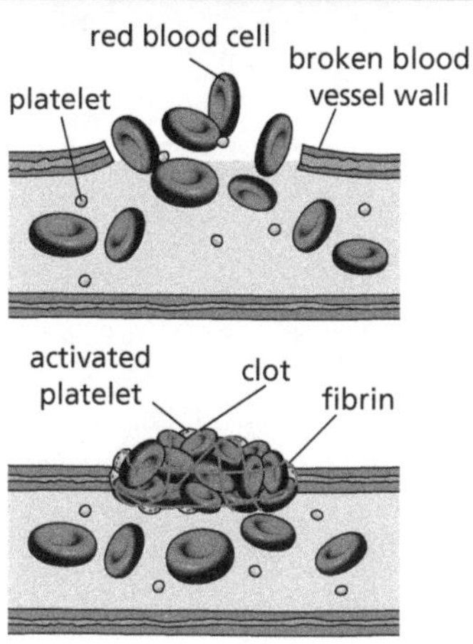

B7.2

The immune system

The **immune system** has two parts. 1. Barriers, which stop pathogens entering the body. 2. White blood cells which attack pathogens if they get into the body.

Barriers like the skin, hair and mucus in the nose, and stomach acid stop pathogens entering the body. If your skin is damaged, the blood quickly clots to reform the skin barrier.

B7.2

The immune system

Collins
Secure Science

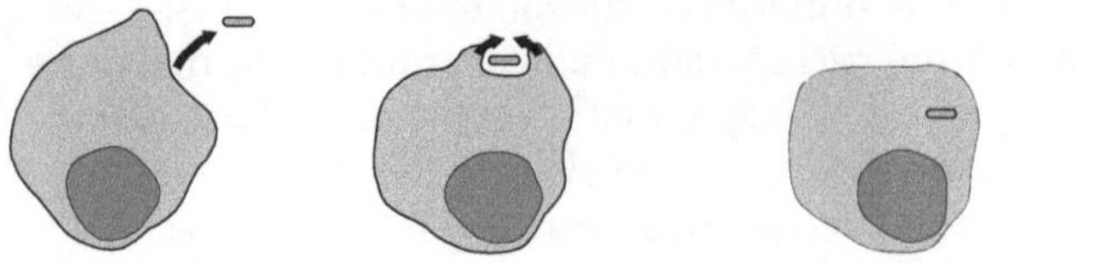

1 A phagocyte moves towards a bacterium.

2 The phagocyte pushes a sleeve of cytoplasm outwards to surround the bacterium.

3 The bacterium is now enclosed in a vacuole inside the cell. It is then killed and digested by enzymes.

B7.3

The immune system

Phagocytes are a type of white blood cell which ingest many different types of pathogen.

The immune system

Collins
Secure Science

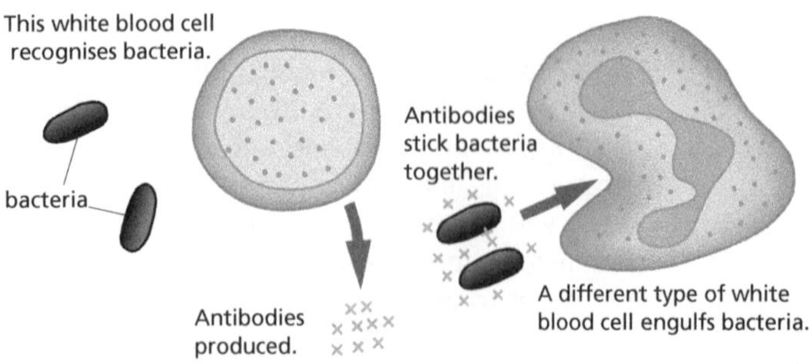

B7.4

The immune system

Other white blood cells called **lymphocytes** are **specific** to a particular pathogen. Lymphocytes make **antibodies** which stick bacteria together and **antitoxins** which neutralise toxins (poisons) made by pathogens.

Pathogens

Collins
Secure Science

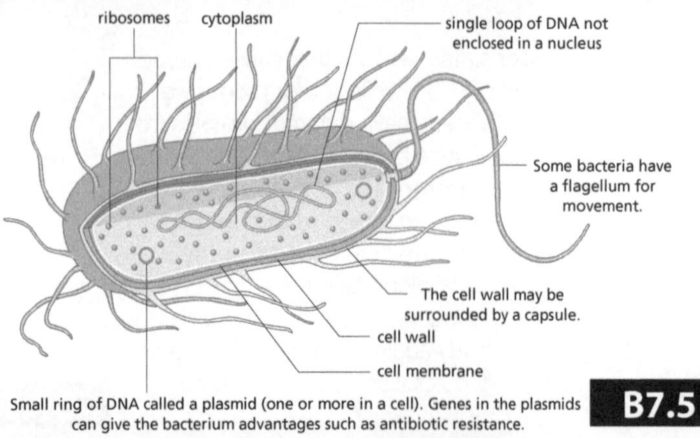

B7.5

Pathogens

Pathogens are organisms or viruses that cause a disease. Pathogens can be **eukaryotes** like fungi and protists, or **prokaryotes** like bacteria, or viruses.

Bacteria are prokaryotic cells which are much smaller than human cells and can be pathogenic. They have different proteins and our body recognises any cell that is not a human cell and makes antibodies to attach to them.

Viruses are very small pathogens which replicate (make copies) inside host cells to make new copies of themselves to be released. This causes damage to the body cells.

Pathogens and their diseases

Collins
Secure Science

Pathogen	Type	Symptoms	How it spreads
Measles	virus	Fever and red skin rash, can be fatal	Inhalation of droplets
HIV	virus	Flu-like illness and eventually an ineffective immune system	Through body fluids like blood and semen
Tobacco mosaic virus	virus	Mosaic pattern of discolouration on plants, decreasing photosynthesis	Direct contact
Salmonella	bacterium	Vomiting, diarrhoea and abdominal cramps	Eating food infected with the bacterium
Gonorrhoea	bacterium	Thick yellow or green discharge from the vagina or penis	Sexual contact
Rose black spot	fungus	Purple or black spots develop on leaves, decreasing photosynthesis	Carried by wind or water
Malaria	protist	Fever, can be fatal	Carried by mosquitoes

B7.6

Preventing disease

Collins Secure Science

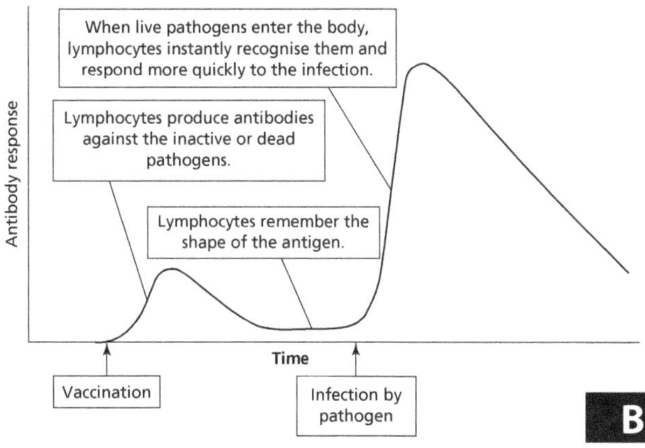

B7.7

Preventing disease

Collins Secure Science

Antibiotics are medicines like **penicillin** which can be given to patients with a bacterial infection. Antibiotics cannot kill **viruses**. Some bacteria are building resistance which means the antibiotics can no longer kill them. **Vaccines** can be used to prevent infection by giving someone small quantities of dead or inactive forms of a pathogen. This stimulates an **immune response**. An immune response means the body quickly makes antibodies if an infection occurs. Vaccines can be used for many different types of pathogen, including viruses.

B7.7

Life cycles

Collins Secure Science

Life cycles

B8.1

Life cycles

Collins Secure Science

All the substances that living things are made of are recycled. Plants fix carbon dioxide from the air and make it into sugar using light in photosynthesis. The sugar is used to build plant bodies. Animals use the carbon fixed by plants to build their own bodies. Animals release energy from sugar from plants using **respiration**. When organisms die, the substances in their bodies are **decomposed** by **microorganisms** so they can be reused.

B8.1

Photosynthesis

Collins Secure Science

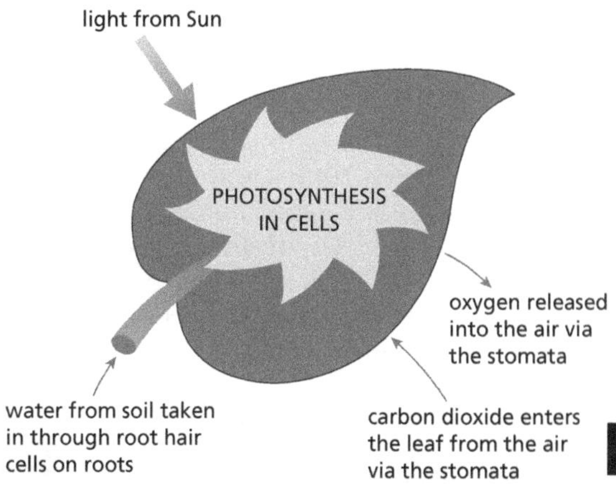

B8.2

Photosynthesis

Collins Secure Science

Photosynthesis is one of the most important chemical reactions in living things as it is the basis for most **food chains**. Plants use the energy from sunlight to turn water (H_2O) and carbon dioxide (CO_2) into simple sugars such as glucose ($C_6H_{12}O_6$).

$$\text{water} + \text{carbon dioxide} \xrightarrow{\text{light and chlorophyll}} \text{glucose} + \text{oxygen (O)}$$

B8.2

Photosynthesis

Collins Secure Science

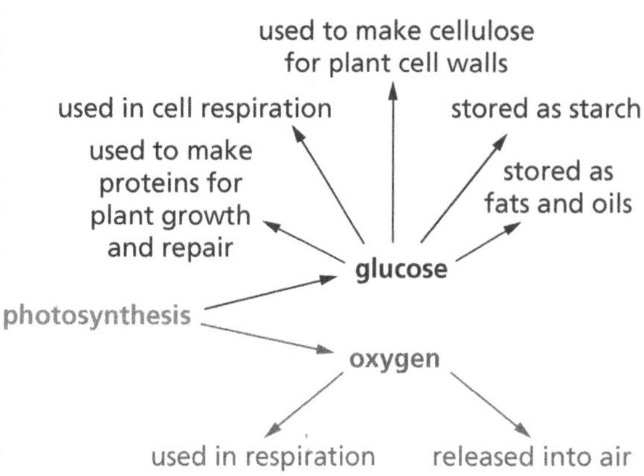

B8.3

Photosynthesis

Collins Secure Science

The sugars produced are stored or used by the plant to grow. Plants also use the sugar in cell respiration along with the oxygen produced in photosynthesis.

B8.3

Nutrient cycling

Collins
Secure Science

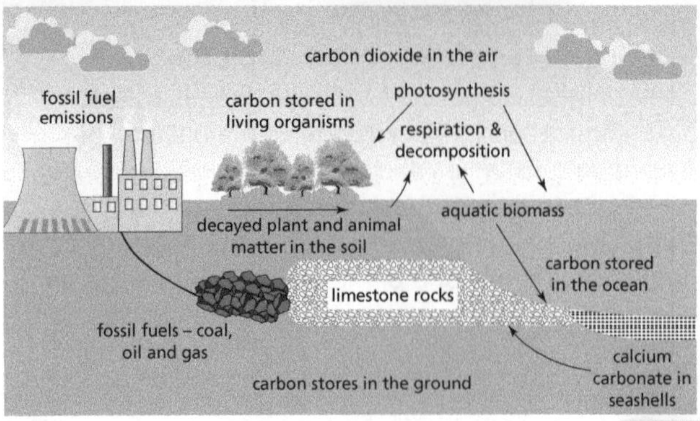

B8.4

Nutrient cycling

Coll
Secure S

Nutrients cycle through the biotic and abiotic parts of an ecosystem through feeding and decomposition. Nutrients in the bodies of organisms are recycled by bacteria and other decomposers in the soil. The nutrients released are available for other organisms to use. Carbon found in living organisms is released to the atmosphere by cell respiration. Carbon is then returned to the food chain through photosynthesis in plants.

The water cycle provides fresh water for animals and plants on land where it eventually drains into the sea. Water evaporates back into clouds and is then returned to the land through **precipitation** (rainfall).

B8.

Human interactions

Collins
Secure Science

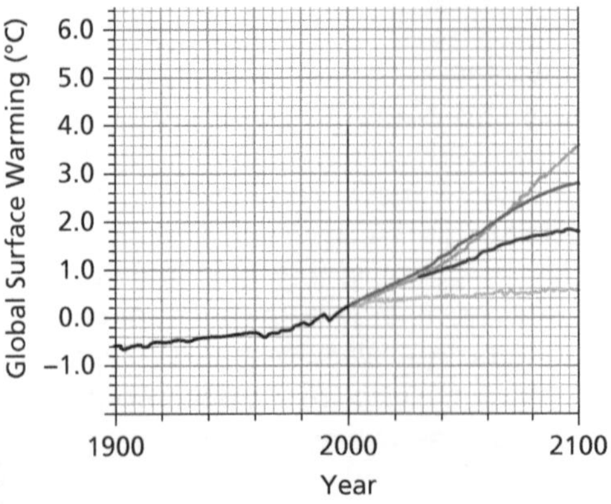

B8.5

Human interactions

Coll
Secure S

Humans have both positive and negative interactions with the environment. Human land use, such as building, quarrying, farming and waste dumping, has led to **deforestation** and habitat destruct The increased release of carbon dioxide and methane into the environment has led to **global warming**. This has resulted in a clim crisis which may result in increased temperatures, sea level rise and of biodiversity. Scientists make predictions about how much the Ear will warm (the graph shows four different models of global warming and these are evaluated using **peer review**. Population growth has led to an increased need for resources which has increased the amo of pollution released. However, humans are now taking action to reduce these negative impacts on biodiversity. Scientists and concern citizens have started breeding programmes for **endangered species**, and encouraged protection and regeneration of rare habitats.

B8

Ecosystems

Collins
Secure Science

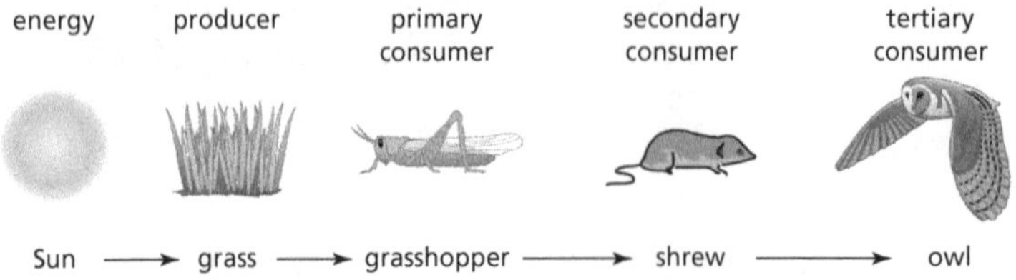

B8.6

Ecosystems

Coll
Secure S

Ecosystems are a web of interactions between living organisms (**biotic factors**) and the **abiotic** (non-living) parts of their surroundings. Within an ecosystem living things **compete** for resources like space and food, but also depend on other species for food and shelter. This is **interdependence**. Some organisms need other species to help them reproduce. For example, flowers need bees for **pollination**. **Biodiversity** is the variety of all the **species** in an ecosystem. Changes to part of an ecosystem can affect all other parts.

B8

Respiration

Collins
Secure Science

Respiration happens in cells. Animals use the sugar from their food, and oxygen for **aerobic respiration**. The equation is:

glucose + oxygen → carbon dioxide + water

B8.7

Respiration

Coll
Secure S

Without oxygen, **anaerobic respiration** takes place in musc

glucose → lactic acid

This does not give as much energy as aerobic respiration.

In plants and yeast, the equation for anaerobic respiration is:

glucose → ethanol + carbon dioxide

Anaerobic respiration in yeast cells is called **fermentation**. Fermentation is used in the making of bread and alcoholic drinks.

Energy transferred by respiration supplies all the energy needed for living processes.

B8

Reproduction and inheritance

Collins
Secure Science

Reproduction and inheritance

B9.1

Reproduction and inheritance

Collins
Secure Science

Characteristics are caused by the code in **DNA**. In eukaryotes DNA is stored in the nucleus of cells. When plants and animals reproduce they make gametes (sex cells) that have half the DNA of body cells. The DNA is different in each gamete and is chosen randomly. DNA can change by **mutations** when cells divide. These processes mean that individuals have lots of small differences from each other. Some of the differences help organisms survive.

B9.1

Meiosis

Collins
Secure Science

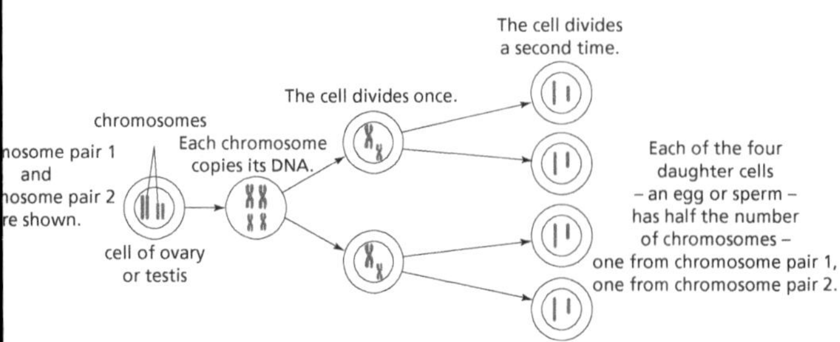

B9.2

Meiosis

Collins
Secure Science

Meiosis is a type of cell division in reproductive organs that makes four **gametes** (sex cells). Gametes are eggs and sperm in humans, and pollen and egg cells in flowering plants. The original cell with chromosomes in pairs divides twice to form four gametes. Each gamete has a single set of chromosomes and is genetically different (non-identical) to the other gametes produced. Meiosis is a way of introducing **variation** into a species.

B9.2

Evolution by natural selection

Collins
Secure Science

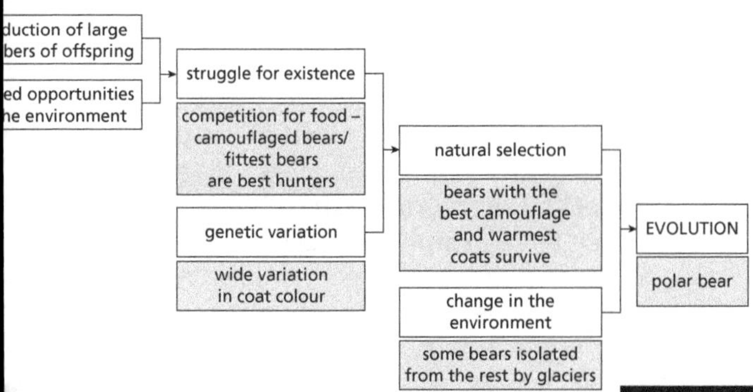

B9.3

Evolution by natural selection

Collins
Secure Science

Evolution is the change in the inherited characteristics of a population over time through **natural selection**. When there is a change in the environment, variation in a species means some individuals are better **adapted** to the new conditions. These individuals therefore survive and reproduce. They pass on the favourable characteristics. Over time the species changes. If two groups of a species are isolated in two different environments natural selection will lead to different adaptations being selected. Eventually, the species will differ so much they become different species and can no longer **interbreed**. The flow chart shows how polar bears may have evolved from brown bears as they were separated by ice.

Fossils provide evidence for evolution by showing changes in organisms that lived before in different aged rocks. Artificial selection (picking individuals with specific traits to breed together) has shown how species can change.

B9.3

Variation

Collins
Secure Science

Darker moth

Light moth

B9.4

Variation

Collins
Secure Science

Variation is important to ensure a species survives changes in the environment. Some individuals are better suited to the new environment. Genetic variation comes from mutation, meiosis and sexual reproduction. Many mutations in the DNA may have no effect. Some mutations could result in a new **phenotype** (characteristic) being produced, such as peppered moths having darker wings.

B9.4

Sexual reproduction

Collins
Secure Science

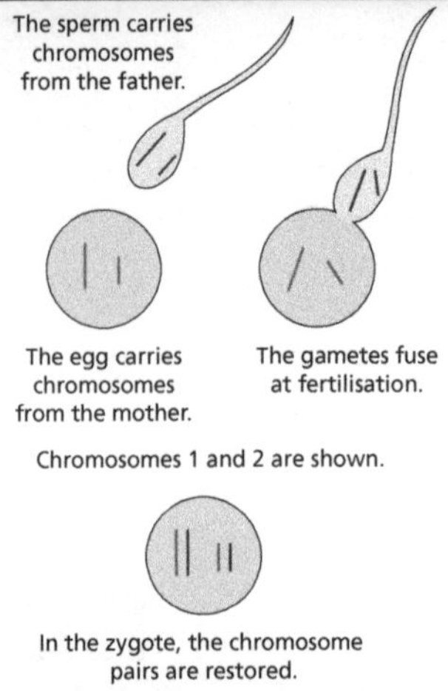

		Mother (XX) gametes	
		X	X
Father (XY) gametes	X	XX female	XX female
	Y	XY male	XY male

B9.5

Sexual reproduction

Coll
Secure S

At **fertilisation** the gametes (sex cells) from each parent fuse to make a fertilised egg. The sex of the baby is determined by the **sex chromosomes**. Egg cells will always have an X chromosome, and sperm cells can have either an X or a Y chromosome. If a sperm with an X chromosome fuses with the egg the baby born will be female with XX sex chromosomes. If the sperm has a Y chromosome the baby born will be male with XY sex chromosomes.

Single-celled organisms can reproduce by **asexual reproduction** but this does not produce variation.

B9

Genetics and genetic engineering

Collins
Secure Science

Genetics and genetic engineering

B10.1

Genetics and genetic engineering

Coll
Secure S

Characteristics are caused by the code in DNA; the form and function of bodies or organisms are controlled by the code of the genetic material, such as DNA. In eukaryotes DNA is stored in the **nucleus** of cells. When plants and animals reproduce they make gametes (sex cells) that have half the DNA of body cells. The DNA is different in each gamete and is chosen randomly. Exact copies of cells are made by division when plants and animals grow. This is called **mitosis**. Humans use **genetic engineering** to make products, plants and animals for food and medicine.

B1(

Mitosis and stem cells

Collins
Secure Science

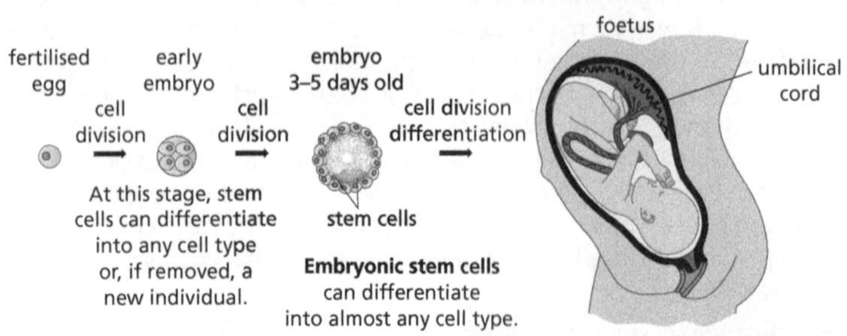

B10.2

Mitosis and stem cells

Coll
Secure S

Cells divide during the stage of the **cell cycle** called mitosis. First the genetic material doubles and then it separates into two identical cells. Mitosis provides cells for growth and repair. **Undifferentiated** cells that can develop into a number of cells are called **stem cells**. **Embryonic stem cells** can produce any type of cell. **Adult stem cells** are more limited and can only produce a small number of different cell types. In **therapeutic cloning** stem cells with the same genes as the patient are used to replace damaged tissues.

B1(

Genes

Collins
Secure Science

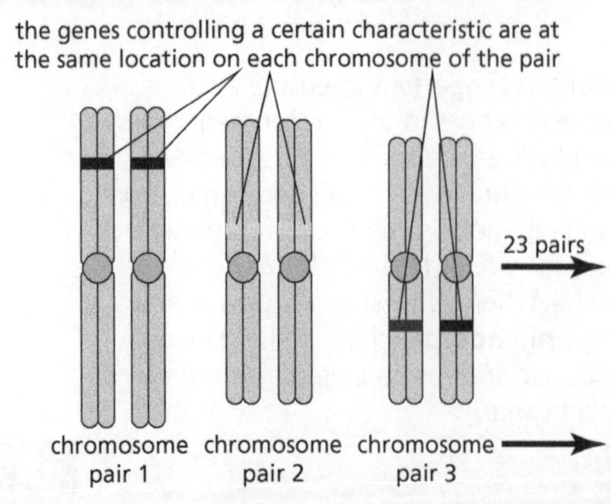

B10.3

Genes

Col
Secure S

Humans have 23 pairs of chromosomes which carry versions of genes called **alleles**. If someone has two copies of the same allele they are **homozygous** for that gene. If they have two different alleles, they are **heterozygous** for that gene. The alleles someone has is their **genotype** and the outward characteristics of their genotype is their phenotype.

B1(

Genetics

		Mother (Ff) gametes	
		F	f
Father (Ff) gametes	F	FF unaffected	Ff unaffected (carrier)
	f	Ff unaffected (carrier)	ff cystic fibrosis

B10.4

Genetics

Dominant alleles are always expressed even if someone only has one copy, like **polydactyly** (a trait where people have extra fingers or toes). **Recessive** alleles can only be expressed if someone has two copies, as in **cystic fibrosis**. You can use a **Punnett square** diagram to predict the **probability** (chance) of the phenotypes of children. You will need to know the genotype of the parents. In the example, both parents are heterozygous (Ff) so they have a 25% chance of having a child with cystic fibrosis (ff).

B10.4

Genetic engineering

For thousands of years people have bred animals or plants which had useful traits. This is called **selective breeding**. After many generations of selective breeding the useful trait is found in all of the offspring. Genetic engineering is a much more direct way and allows the addition of useful genes from one species into another.

B10.5

Genetic engineering

Genetically modified or **GM crops** are formed by genetic engineering. This has allowed farmers to grow plants that give bigger **yields** and are resistant to disease and attack from insects. Some people are worried about potential harmful effects to human health and the environment. Genetic engineering might also be used to treat some inherited disorders like cystic fibrosis.

B10.5

DNA

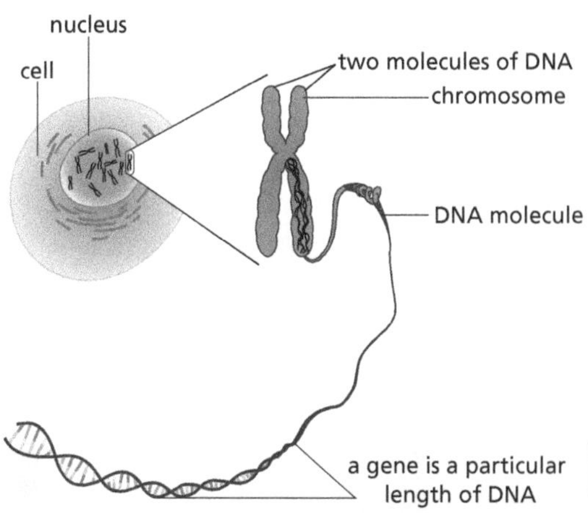

B10.6

DNA

The **genome** (all the genetic material) of a human cell is made of DNA, which is in the shape of a **double helix**. The DNA coils up to form **chromosomes**. A small section of DNA is called a **gene**. Each gene carries the code for a specific protein.

B10.6

Making a slide

1. Add a few drops of water onto the slide using a pipette.
2. Place the sample flat on the surface of the slide using tweezers.
3. Add one or two drops of iodine solution to the sample from a dropping bottle to stain certain structures and allow them to be seen more clearly.
4. Using a mounted needle or seeker, lower the coverslip onto the slide carefully to make sure there are no air bubbles.

B11.1

Using a microscope

Collins
Secure Science

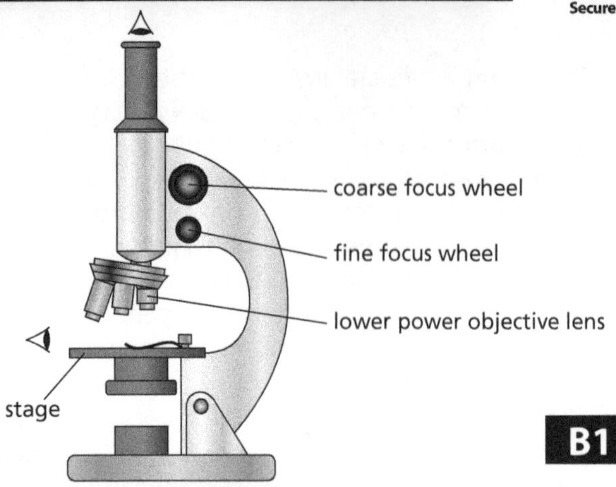

B11.2

Using a microscope

Coll
Secure S

1. Select the lowest power objective lens.
2. Turn on the light and/or adjust the mirror so light is visible through the eyepiece lens.
3. Place the prepared slide onto the stage and secure with the stage clips.
4. Use the coarse focus wheel to bring the stage and objective lens as close to each other as possible without the lens touching the slide.
5. Look through the eyepiece and turn the coarse focus wheel so that the eyepiece lens and sample move apart, until the image comes into view.
6. Use the fine focus wheel to focus the image.

B11

Investigating osmosis

Collins
Secure Science

1. Cut plant tissue pieces to the same length and shape. Note down the starting mass for each piece.
2. Place the pieces of plant tissue into different solutions of salt.
3. Wait at least 10 minutes. Dry the pieces and then measure the mass of each piece of tissue.
4. Calculate the percentage change in mass:

$$\text{percentage change in mass} = \frac{(\text{final mass} - \text{starting mass})}{\text{starting mass}} \times 100$$

B11.3

Investigating molecules in food

Collins
Secure Science

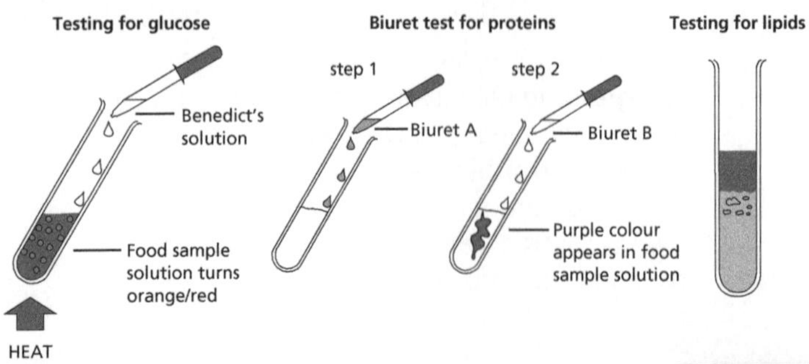

B11.4

Investigating molecules in food

Coll
Secure S

1. Make food solutions by grinding some food to a paste and adding some to a test tube.
2. To test for sugar: add Benedict's solution and then place in a water bath of boiling water.
3. A change to orange/red indicates that sugar is present.
4. To test for protein: first add Biuret A solution and then Biuret B solution.
5. A change to purple indicates that protein is present.
6. To test for lipids: filter some of the food solution into a test tube. Add Sudan III stain and shake gently.
7. A change to red/orange indicates that lipids are present.

B1

Calculating total magnification

Collins
Secure Science

Diagrams of magnified cells need to include the total magnification. If the eyepiece has a magnification of 10, which we write as ×10, and the objective lens has a magnification of ×40, then in total the cell has been magnified by ×400.

$$\text{total magnification} = \text{eyepiece lens magnification} \times \text{objective lens magnification}$$

B11.5

Unit conversion

Collins
Secure Science

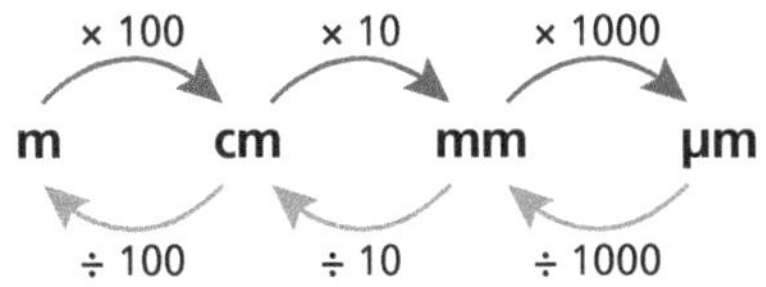

B11.6

alculating magnification of an image

Collins
Secure Science

$$\text{magnification of image} = \frac{\text{size of image}}{\text{size of real object}}$$

Sometimes you may be told the size of the image; sometimes you may need to measure it with a ruler yourself.

Make sure that the size of the image and the size of the real object are in the same unit, and convert if you need to.

Then divide the size of image by the size of the real object to calculate the magnification.

B11.7

Investigating rate of reaction

Collins
Secure Science

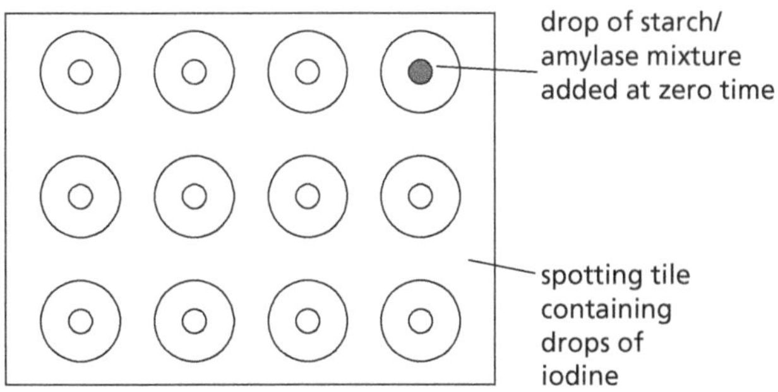

B11.8

Investigating rate of reaction

Collins
Secure Science

The enzyme amylase breaks down starch. Iodine turns blue/black in the presence of starch.

1. Add a drop of iodine to each well.
2. In a test tube, mix starch and amylase.
3. Add a drop of starch/amylase mixture to the first well – this represents zero time.
4. After 30 seconds add another drop of the starch/amylase mixture to well 2.
5. Repeat every 30 seconds in wells 3, 4, etc., to show how long it takes for mixture to go from blue/black to brown.
6. Repeat steps 1–5 with different pH buffers to investigate the effects of pH.

B11.8

Measuring population size

Collins
Secure Science

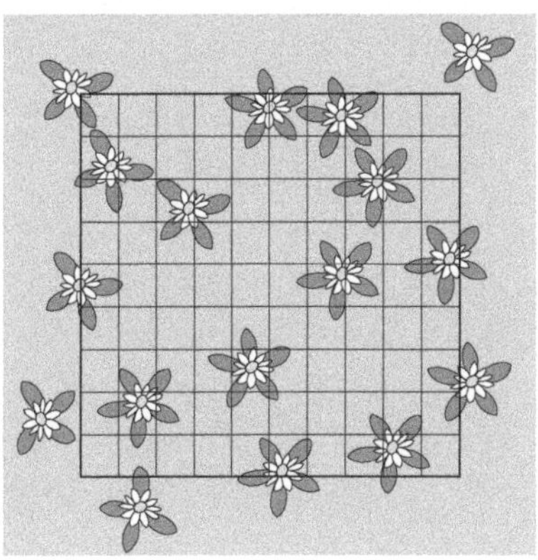

B11.9

Measuring population size

Collins
Secure Science

1. Choose which plant or plants you want to investigate.
2. Decide whether it is best to measure the number of plants or percentage cover.
3. Decide how to make your sample **random** and **representative**.
4. Place the **quadrat** down and count the presence of your plant or plants.
5. Keep repeating until your sample is representative.
6. Calculate the area covered by the plant:
   - Count each square that the plant covers in your quadrat.
   - Decide whether you will count only squares completely covered or also include partially covered squares.
   - To calculate the area of a quadrat multiply the height by the width (e.g. 100 cm × 100 cm = 10 000 cm^2 or 1 m^2.
   - To work out how many of your chosen plants there are in the habitat you are investigating: (1) divide the total area by the size of your quadrat; (2) multiply the average number of your plant per quadrat by the number of quadrats that fit in your habitat.

B11.9

Ethics in biology experiment

Collins
Secure Science

1. If you are using people in your experiment, e.g. measuring reaction times, get informed consent before doing the experiment.
2. If you are using plant samples, e.g. investigating light, only pick a sustainable sample so that the plant can recover.
3. If you are measuring your sample in the environment, e.g. measuring population size, do not damage any plants or animals that you are observing.

B11.10

Investigating reaction time

Collins
Secure Science

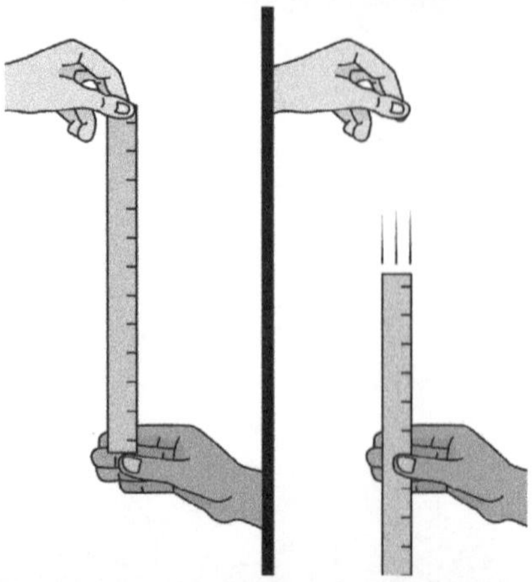

B11.11

Investigating reaction time

Col
Secure S

1. In a pair take it in turns for one student to drop a 30 cm ruler. The other student then catches the ruler between their thumb and forefinger.
2. Measure the distance the ruler drops.
3. Use a conversion table to convert distance of ruler drop into reaction time.
4. Repeat steps 1–3 to test the effects of a factor such as different levels of background noise.

B11

Investigating photosynthesis

Collins
Secure Science

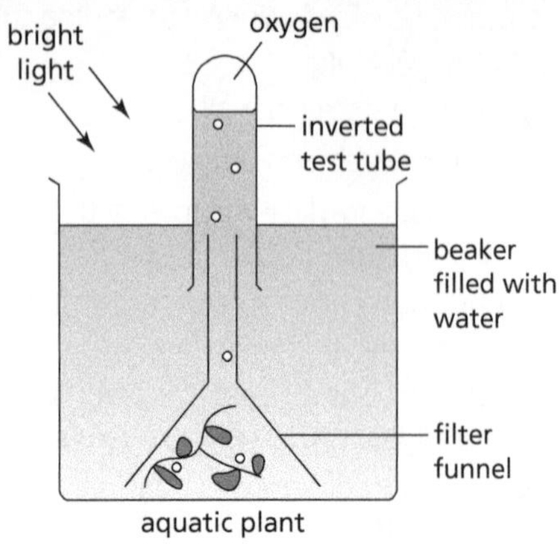

B11.12

Investigating photosynthesis

Col
Secure S

1. Place an aquatic plant in a beaker of pond water under a funnel with an upturned test tube.
2. Place a lamp next to the beaker and measure how much oxygen from photosynthesis is collected in the test tube.
3. Move the lamp different distances from the plant – measure how the oxygen production changes.

B11

The particle model

C12.1

Session 12: Knowledge organiser
Collins Secure Science

The **particle model** helps us to understand the differences between solids, liquids and gases. The **properties** of a material are determined by how particles are arranged. In a solid, particles can only vibrate in a fixed pattern. In liquids, particles can move over each other but are always touching other particles. Particles in a gas do not touch each other and there is lots of space between particles.

C12.1

Particle model

Collins Secure Science

Everything on Earth is made of matter that exists in three different states: solid, liquid, gas. The particle model states that matter consists of very tiny particles that are constantly moving. This model is not perfect, but it helps us to explain the properties of solids, liquids and gases. In reality, the gas particles are much further apart, and the particles are not all the same size.

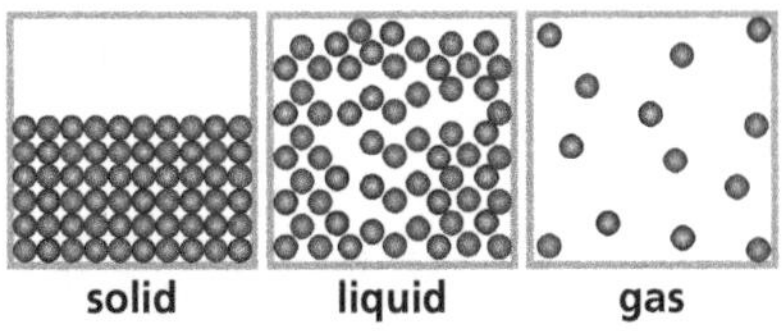

C12.2

Particle model

Collins Secure Science

In solids, the particles are held together by strong forces of attraction, called bonds. A solid has a fixed size and shape. In liquids, the forces of attraction between the particles are weaker than in solids. There is no fixed structure. In gases the particles are almost free of any attractions between them. They move randomly and quickly, colliding with each other and the walls of their container.

C12.2

Density

Collins Secure Science

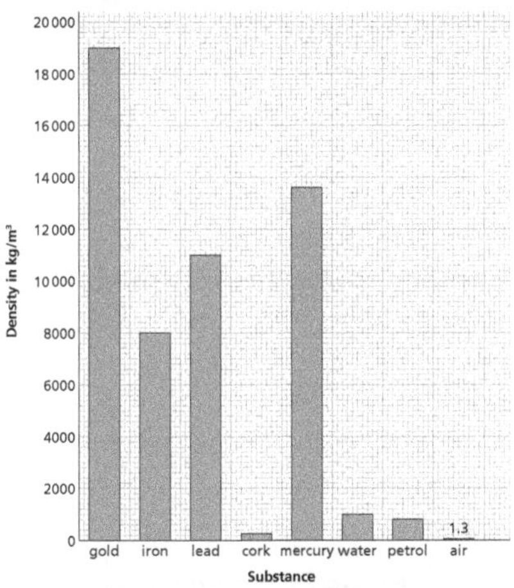

C12.3

Density

Collins Secure Science

Density compares the mass of materials that have the same volume. It is calculated from the equation

$$\text{density (kg/m}^3\text{)} = \frac{\text{mass (kg)}}{\text{volume (m}^3\text{)}}$$

The density of a material depends on the size of the particles and how closely they are packed. So most liquids are less dense than solids.

C12.3

Internal energy

Collins Secure Science

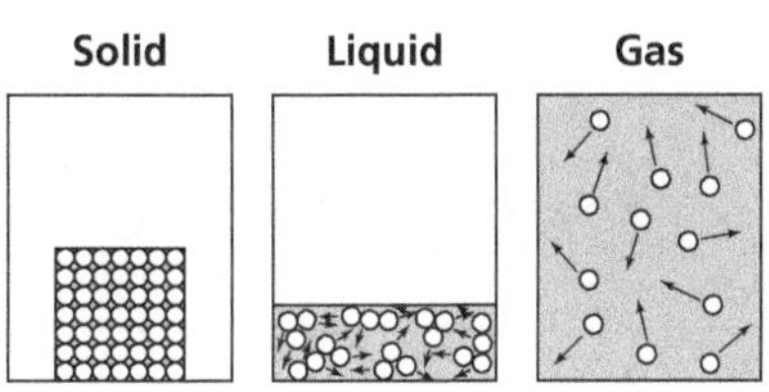

C12.4

Internal energy

Collins Secure Science

Particles are always moving. They have **kinetic energy**. At the same temperature, all particles have the same average kinetic energy. Therefore, heavy particles move more slowly than lighter particles.

Particles also have **potential energy**. The **internal energy** of a system is the total kinetic and potential energy of all the particles in the system. The amount of energy stored in or released from a system as the temperature changes can be calculated using the equation:

change in thermal energy = mass (kg) × specific heat capacity (J/kg °C) × temperature change (°C)

$$\Delta E = m\, c\, \Delta\theta$$

C12.4

Changing state

Collins
Secure Science

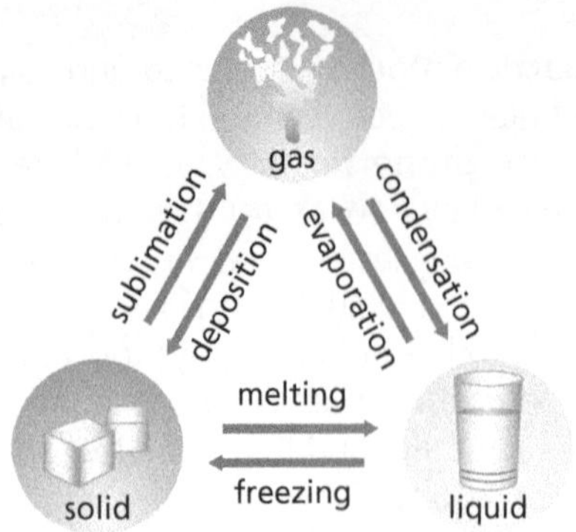

C12.5

Changing state

Coll
Secure

We can change the state of matter of any substance by heating or cooling. This will change the internal energy of a substance. As **thermal energy** is added to a solid, the particle vibrations increase and the forces holding the particles in place are stretched and pushed further apart. Eventually, some of the forces break and the particles can move over each other. The substance is now in the liquid state. This process is **melting**. When even more thermal energy is added, the remaining forces break and the particles are free to move from place to place. Changes of state are physical changes which differ from chemical changes because the material recovers its original properties if the change is reversed.

C12

Changing state

Collins
Secure Science

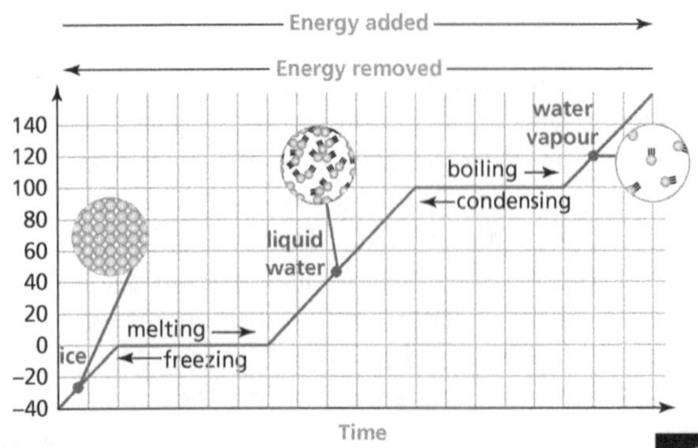

C12.6

Changing state

Coll
Secure

When you heat a substance there are stages where the heat added causes a rise in temperature. Then, as the substance changes state, the temperature stops rising because the heat is used to break the forces.

C12

Session 13: Atomic structure

Collins
Secure Science

Atomic structure

C13.1

Session 13: Atomic structure

Coll
Secure

Atoms are made of **neutrons**, **protons** and **electrons**. Each element has a specific number of protons in the **nucleus** and the same number of electrons in shells around the nucleus. The **periodic table** lists the number of protons of the elements and the symbol for each element. The periodic table is arranged so that you can see patterns in the types and strength of the reactions of each element relative to each other.

C13

Atomic structure

Collins
Secure Science

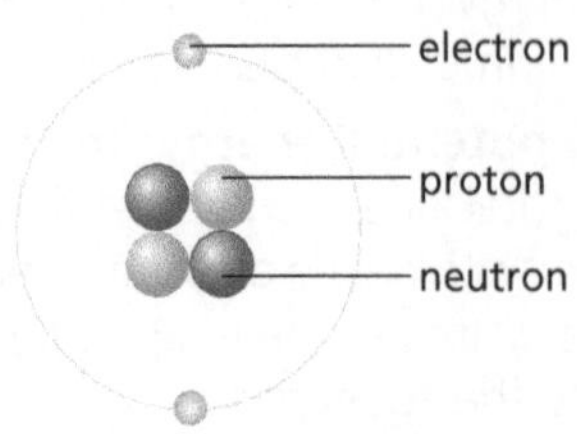

C13.2

Atomic structure

Coll
Secure

The nucleus of a helium atom contains two positively charged protons and two neutrons. It is surrounded by two negatively charged electrons. The relative masses and charges of the particles are shown in the table.

	Relative charge	Relative mass
Electron	–1	0.0005
Proton	+1	1
Neutron	0	1

C1

Atomic structure

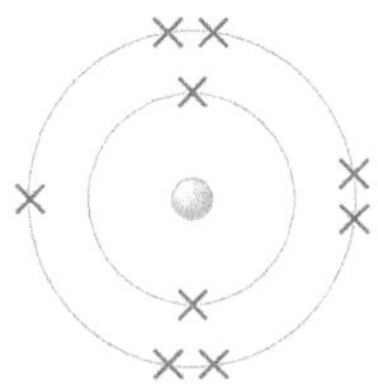

Atomic structure

Collins
Secure Science

The electrons in an atom are arranged in energy levels or shells. The first energy level can hold 2 electrons and the second and third levels can hold up to 8 electrons each. This atom of fluorine has 9 electrons: 2 in the first shell and 7 in the second shell. The **electronic structure** can be written 2,7 which makes it easy to find the element in the periodic table. This is because the group number is the same as the number of electrons in the outer shell of all the elements in that group.

C13.3

Periodic table

Collins
Secure Science

Group 2											Group 3	4	5	6	7	0
		$^{1}_{1}\text{H}$ hydrogen														$^{4}_{2}\text{He}$ helium
$^{9}_{4}\text{Be}$ beryllium											$^{11}_{5}\text{B}$ boron	$^{12}_{6}\text{C}$ carbon	$^{14}_{7}\text{N}$ nitrogen	$^{16}_{8}\text{O}$ oxygen	$^{19}_{9}\text{F}$ fluorine	$^{20}_{10}\text{Ne}$ neon
$^{24}_{12}\text{Mg}$ magnesium											$^{27}_{13}\text{Al}$ aluminium	$^{28}_{14}\text{Si}$ silicon	$^{31}_{15}\text{P}$ phosphorus	$^{32}_{16}\text{S}$ sulfur	$^{35}_{17}\text{Cl}$ chlorine	$^{40}_{18}\text{Ar}$ argon
$^{40}_{20}\text{Ca}$ calcium	$^{45}_{21}\text{Sc}$ scandium	$^{48}_{22}\text{Ti}$ titanium	$^{51}_{23}\text{V}$ vanadium	$^{52}_{24}\text{Cr}$ chromium	$^{55}_{25}\text{Mn}$ manganese	$^{56}_{26}\text{Fe}$ iron	$^{59}_{27}\text{Co}$ cobalt	$^{59}_{28}\text{Ni}$ nickel	$^{64}_{29}\text{Cu}$ copper	$^{65}_{30}\text{Zn}$ zinc	$^{70}_{31}\text{Ga}$ gallium	$^{73}_{32}\text{Ge}$ germanium	$^{75}_{33}\text{As}$ arsenic	$^{79}_{34}\text{Se}$ selenium	$^{80}_{35}\text{Br}$ bromine	$^{84}_{36}\text{Kr}$ krypton
$^{88}_{38}\text{Sr}$ strontium	$^{89}_{39}\text{Y}$ yttrium	$^{91}_{40}\text{Zr}$ zirconium	$^{93}_{41}\text{Nb}$ niobium	$^{96}_{42}\text{Mo}$ molybdenum	$^{99}_{43}\text{Tc}$ technetium	$^{101}_{44}\text{Ru}$ ruthenium	$^{103}_{45}\text{Rh}$ rhodium	$^{106}_{46}\text{Pd}$ palladium	$^{108}_{47}\text{Ag}$ silver	$^{112}_{48}\text{Cd}$ cadmium	$^{115}_{49}\text{In}$ indium	$^{119}_{50}\text{Sn}$ tin	$^{122}_{51}\text{Sb}$ antimony	$^{128}_{52}\text{Te}$ tellurium	$^{127}_{53}\text{I}$ iodine	$^{131}_{54}\text{Xe}$ xenon

Periodic table

Collins
Secure Science

All known elements are listed in the periodic table (part of which is shown here). The metals are found on the left and the non-metals on the right. Each element has its own symbol, with two numbers next to it. If the symbol has more than one letter, then the first letter is always a capital letter. The **atomic number** (bottom number), tells us the number of protons in an atom. The number of electrons is the same as the number of protons. The **mass number** (upper number), tells us the total number of protons and neutrons present in an atom.

The elements are arranged in order of increasing atomic number. The table is divided into periods (rows) and groups (columns). Elements in the first period only have electrons in the first shell. Elements in the third period have electrons in the first three shells. The group number refers to the number of electrons in the outer shell, apart from Group 0, which contains the **noble gases**, where the outer electron shell is full, making them unreactive. The term Group 0 is used to avoid confusion because, although most elements in this group have 8 electrons in their outer shell, helium only has two.

C13.4

Periodicity

Collins
Secure Science

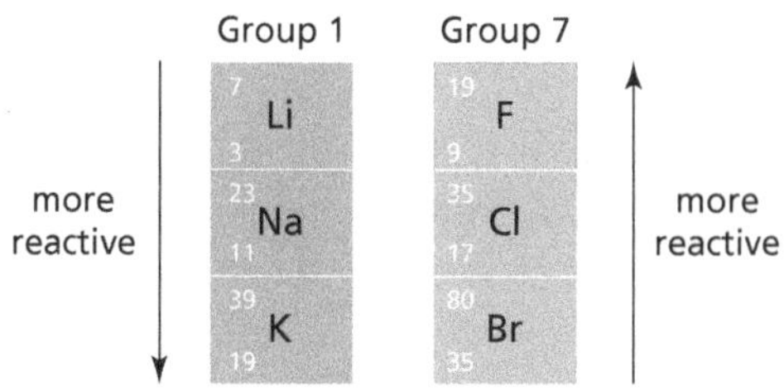

Periodicity

Collins
Secure Science

The trends in physical and chemical properties both across and down the periodic table depend on the atomic structure and the electronic structure. Melting and boiling points of elements in the same group change as you go down the group.

Groups contain elements with similar chemical properties. For example, the elements in Group 1 are all highly reactive because they have 1 electron in their outer shell. The reactivity increases down the group, as the outer electron is further away from the nucleus. Therefore, the **electrostatic attraction** with the nucleus is weaker. The outer electron is 'transferred' during chemical reactions. The reactivity of Group 7 increases as you go up the group because the number of electron shells decreases, making it easier for an electron to be attracted.

C13.5

Formula

Collins
Secure Science

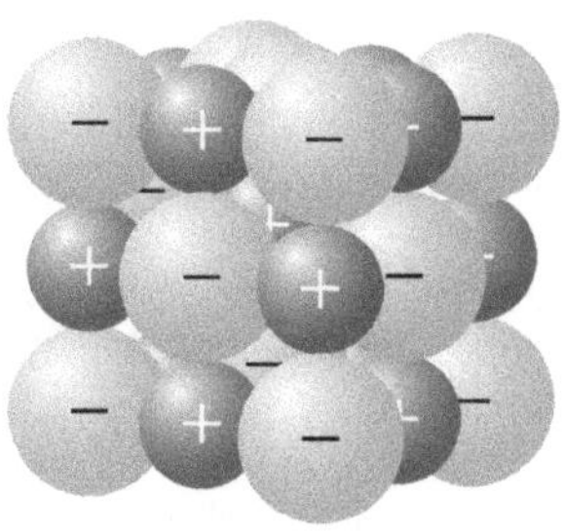

Group 1 element +
Group 7 elements –

Formula

Collins
Secure Science

All substances are chemicals made from atoms. You can see which elements are present by looking at the chemical formula. Elements from Group 1 and Group 7 combine to make compounds containing ions of each element in a fixed ratio of 1:1.

Ions with opposite charges are attracted together to form giant crystal lattices.

C13.6

Formula

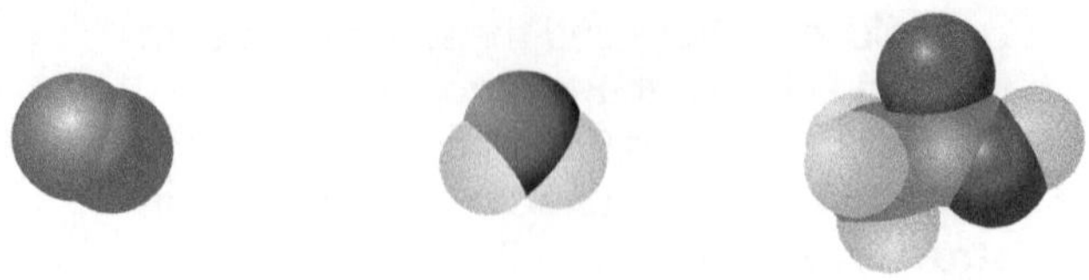

C13.7

Formula

The formula of sodium chloride is NaCl. The formula of magnesium chloride is $MgCl_2$ which tells us that there is one magnesium ion for every two chloride ions present.

The oxygen molecule contains 2 oxygen atoms and is written O_2. The water molecule contains 2 hydrogen atoms and 1 oxygen atom and is written H_2O. Ethanoic acid contains 2 carbon atoms, 2 oxygen atoms and 4 hydrogen atoms and is written CH_3CO_2H.

C13

Session 14: Calculating chemical change

Calculating chemical change

C14.1

Session 14: Calculating chemical change

During a chemical reaction, atoms are rearranged. All the atoms that were there at the beginning of the reaction still exist in the **products**. But the products will have different properties. Energy can be transferred from or to the surroundings during a chemical reaction. Chemical changes can be shown using word equations or symbol equations. As we know that no atoms are lost, we can calculate the amounts of **reactants** and products.

C14

Chemical and physical change

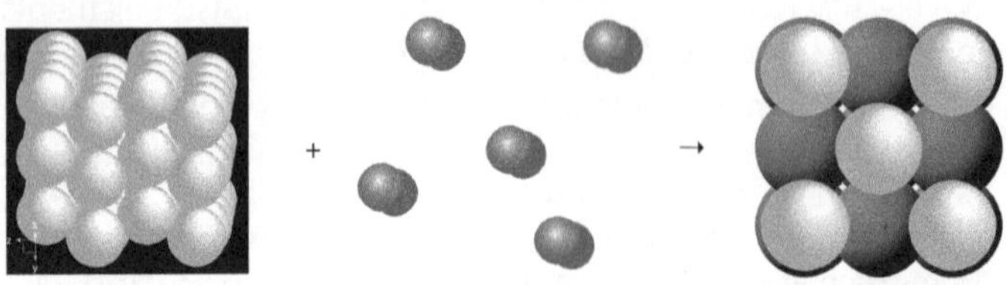

magnesium (grey solid) + oxygen (colourless gas) → magnesium oxide (white solid powder)

C14.2

Chemical and physical change

During a chemical change, substances react together to make new substances with different properties.

When iodine **sublimes** it changes from a solid to a gas. A physical change takes place. The only change to the iodine atoms is an increase in internal energy.

C14

Word and symbol equations

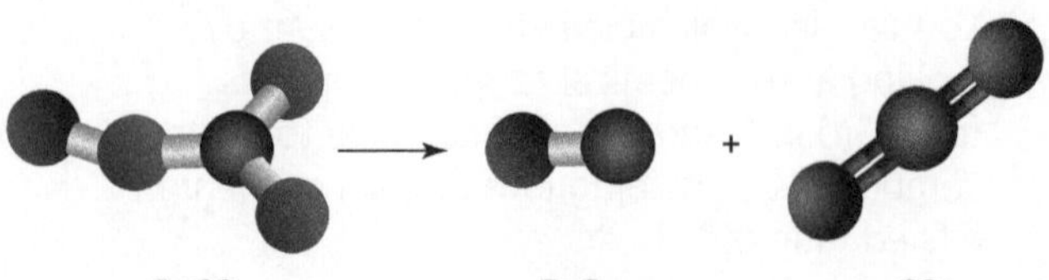

$ZnCO_3$ ⟶ ZnO + CO_2

C14.3

Word and symbol equations

Chemical equations are used to describe what is happening in a chemical reaction.

reactants → products

For example:

zinc carbonate → zinc oxide + carbon dioxide

A word equation gives the names of the reactants (starting substances) products (finishing products). The arrow tells us that a change is taking Symbol equations show the formulae of all the reactants and produc They should be balanced, showing mass is conserved during the reac

$$Ca + O_2 \rightarrow CaO \quad ✗$$

This equation is not balanced. There are 2 oxygen atoms on the left and only 1 on the right side. Each oxygen atom will join to a calcium atom. Therefore, we also need 2 calcium atoms on the left side.

$$2Ca + O_2 \rightarrow 2CaO \quad ✓$$

It is now balanced.

C1

Conservation of mass

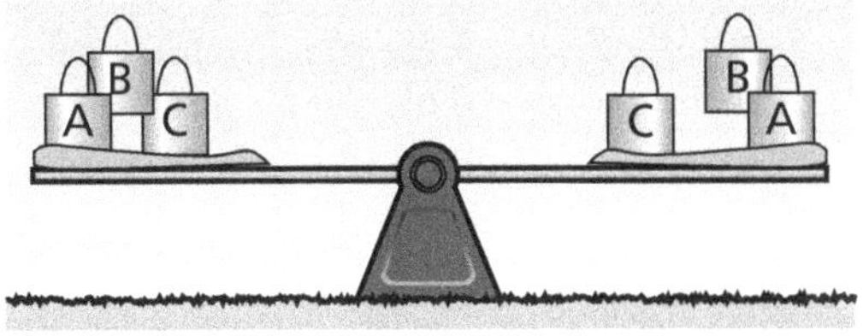

C14.4

Conservation of mass

Collins
Secure Science

When a physical or chemical change takes place, no atoms are made or lost. During a chemical change, atoms are just rearranged. The masses of reactants and products are balanced. This is known as the law of **conservation of mass**. Energy can be transferred to or from the surroundings.

C14.4

Conservation of mass

Collins
Secure Science

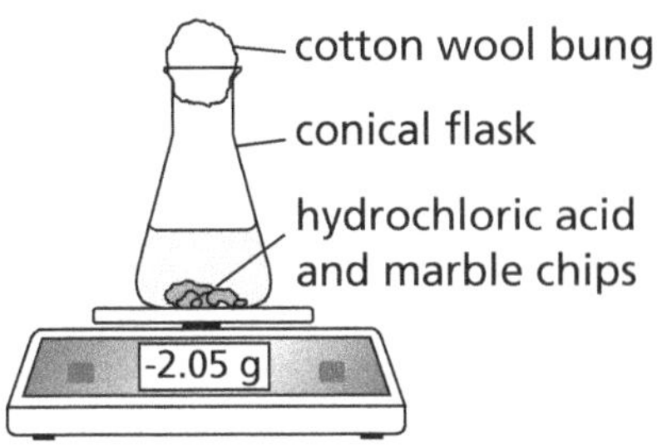

C14.5

Conservation of mass

Collins
Secure Science

Some reactions may appear to involve a change in mass. This can usually be explained because a reactant or product is a gas and its mass has not been taken into account.

In this reaction

magnesium carbonate + hydrochloric acid → magnesium chloride + water + carbon dioxide

the mass appears to decrease. This is because carbon dioxide gas is produced and escapes from the flask into the atmosphere.

C14.5

Calculating chemical quantities

Collins
Secure Science

A_r: Zn = 65; C = 12; O = 16

	$ZnCO_3$	→	**ZnO**	+	**CO_2**
M_r:	65 + 12 + (16 × 3)		65 + 16	+	12 + (16 × 2)
	65 + 12 + 48	→	81	+	12 + 32
	125	→	81	+	44
	125	→	125		

C14.6

Calculating chemical quantities

Collins
Secure Science

In a reaction, how much product will we make? How much reactant is needed? We can calculate the amounts of reactants and products in a chemical reaction using the balanced equation and the **relative formula mass** (M_r).

M_r of a compound is the sum of the **relative atomic masses** (A_r) of the atoms in the numbers shown in the formula. You can take the A_r value as the mass number from the periodic table.

If we start with 12.5 g of $ZnCO_3$, the multiplier is $\frac{12.5}{125} = 0.1$

We will produce 0.1 × 81 = 8.1 g of ZnO and 0.1 × 44 = 4.4 g of CO_2.

C14.6

Test for common gases

Collins
Secure Science

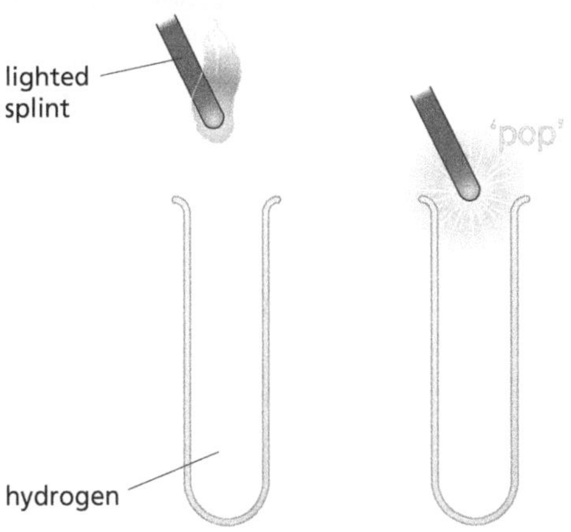

C14.7

Test for common gases

Collins
Secure Science

Test for hydrogen: a burning splint is held at the open end of a test tube containing the gas. Hydrogen burns rapidly with a 'squeaky' pop.

C14.7

Test for common gases

Collins
Secure Science

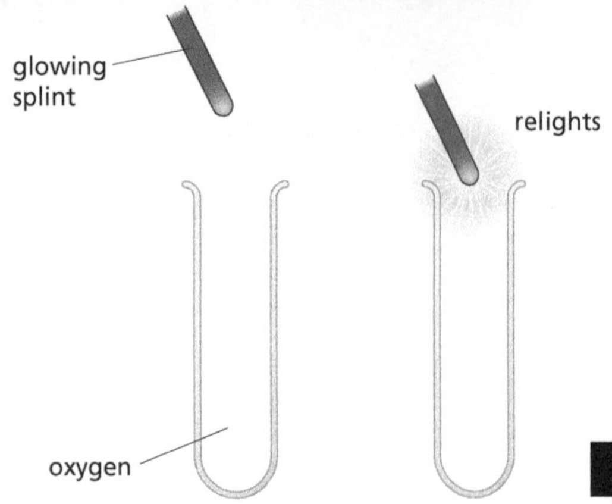

C14.8

Test for common gases

Coll
Secure S

Test for oxygen: a glowing splint is inserted into a test tube containing the gas. The splint relights if the gas is oxygen.

C14

Test for common gases

Collins
Secure Science

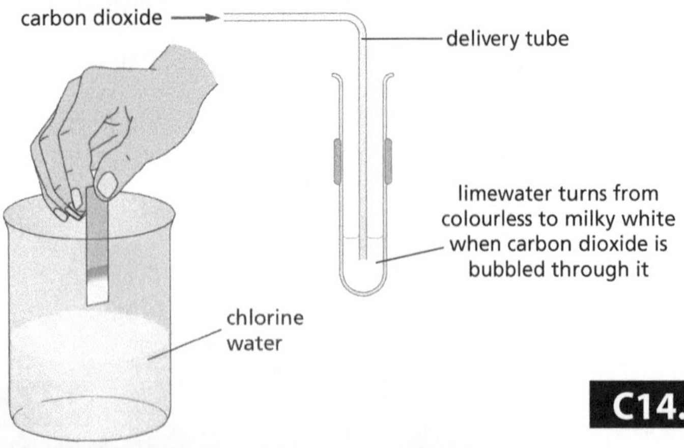

C14.9

Test for common gases

Coll
Secure S

Test for carbon dioxide: the gas is bubbled through a solution of calcium hydroxide (called limewater). Carbon dioxide gas turns the limewater milky (cloudy).

Test for chlorine: damp litmus paper is put into the gas. If the gas is chlorine, it will bleach the litmus paper, turning it white.

C14

Session 15: Bonding

Collins
Secure Science

Bonding

C15.1

Session 15: Bonding

Coll
Secure S

Electrons can be shared with another atom in a **covalent bond**. Electrons can also move between atoms to form ions during the formation of an **ionic bond**. In both cases, atoms share, donate or accept an electron to gain a full outer shell. In metals, the metal ions are surrounded by a sea of electrons donated by the metal atoms. How particles are arranged depends on the strength of the bonds and forces that hold the particles in position.

C15

Ionic bonding

Collins
Secure Science

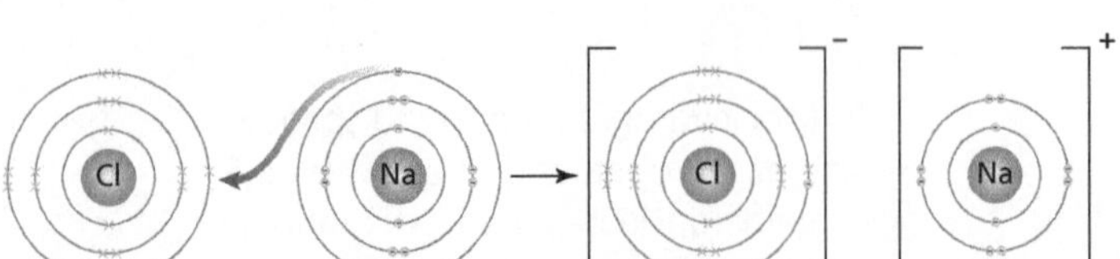

C15.2

Ionic bonding

Coll
Secure S

When a metal atom reacts with a non-metal atom, electrons in the outer shell of the metal atom are transferred. Both positively charge and negatively charged ions are formed. The ions are held together by an electrostatic force of attraction called the ionic bond. The electron transfer during the formation of an **ionic compound** can b represented by a **dot and cross diagram**. The dots represent electr from one atom and the crosses represent electrons from another at

The ions produced have the electronic structure of a noble gas. Sometimes you might see the dot and cross diagram written like thi

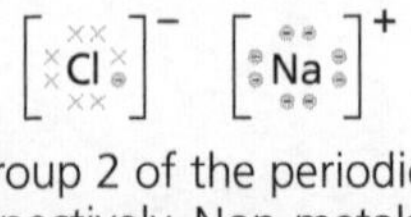

Metals in Group 1 and Group 2 of the periodic table produce ions w a charge of + and 2+, respectively. Non-metals in Group 6 and Group 7 produce ions with a **charge** of 2– and –, respectively. Ionic compounds form **giant crystal lattices**.

C1

Ionic bonding

Collins Secure Science

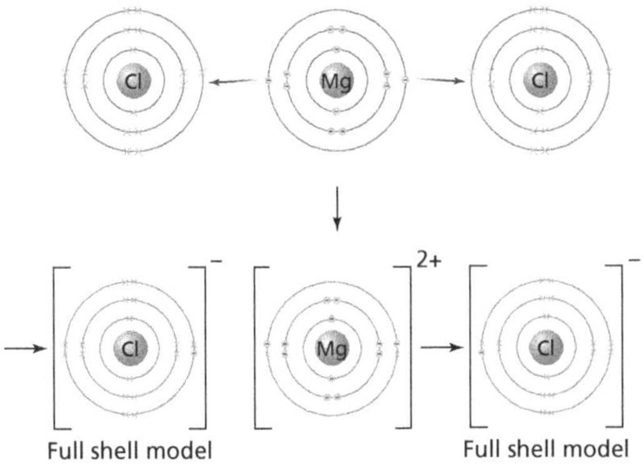

C15.3

Ionic bonding

Collins Secure Science

This diagram shows the formation of magnesium chloride from one magnesium and two chlorine atoms.

C15.3

Type of bonding

Collins Secure Science

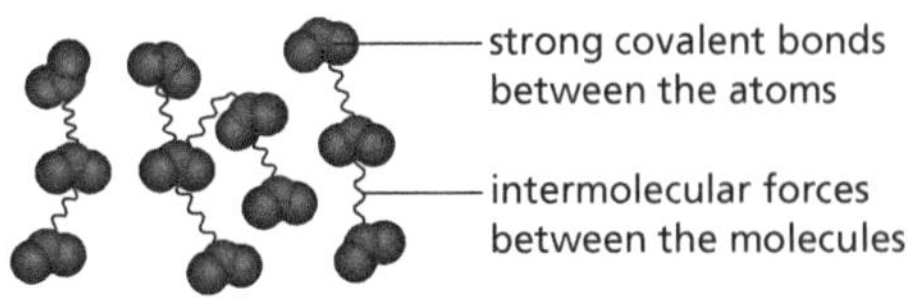

C15.4

Type of bonding

Collins Secure Science

A chemical bond is a force of attraction between different particles. There are two types of bonds found in and between simple molecules such as water (H_2O). A strong covalent bond holds the hydrogen and oxygen atoms together in each molecule. Weaker bonds called **intermolecular forces** exist between the water molecules.

C15.4

Type of bonding

Collins Secure Science

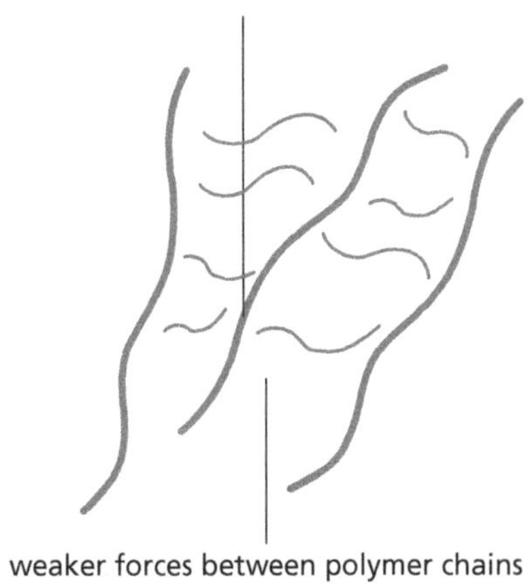

C15.5

Type of bonding

Collins Secure Science

Intermolecular forces play an important role in **polymers**, explaining why polymers can be stretched without breaking. The atoms in the polymer molecules are linked to other atoms by strong covalent bonds to form a long chain. The intermolecular forces between the polymer chains are weaker. They allow the polymer chains to slide over each other but are not so weak to allow the chains to be pulled apart.

C15.5

Metallic bonding

Collins Secure Science

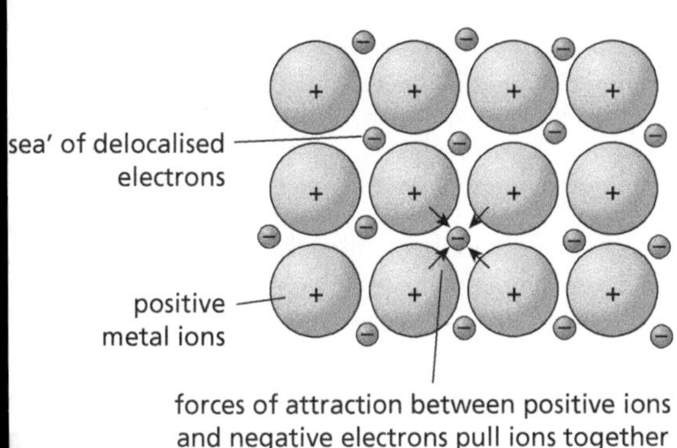

C15.6

Metallic bonding

Collins Secure Science

Metals consist of giant structures of atoms arranged in a regular pattern. Each metal atom has one or more electrons in its outer shell, which are free to move between and around other atoms. These electrons are called **delocalised electrons**. The metal atoms that 'lose' these delocalised electrons become positively charged ions. The **metallic bond** is the strong electrostatic force of attraction between the positive metal ions and the delocalised electrons.

C15.6

Covalent bonding

Collins
Secure Science

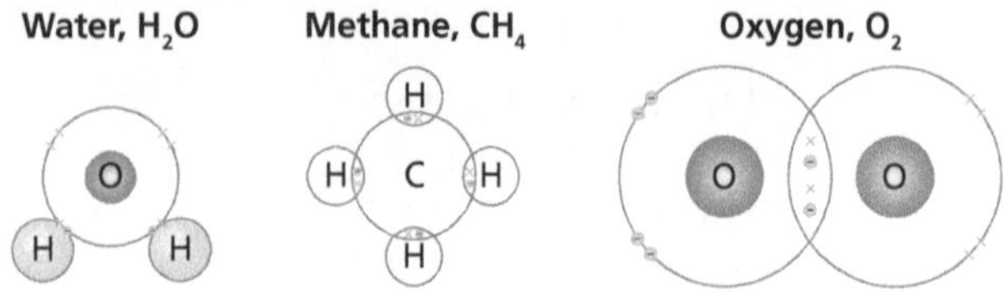

C15.7

Covalent bonding

Coll
Secure

When two or more non-metal atoms bond together, they form a molecule. Covalent molecules are held together by covalent bonds, where electron pairs are shared between two atoms. Dot and cross diagrams are used to represent covalent bonding.

C15

Session 16: Properties of materials

Collins
Secure Science

Properties of materials

C16.1

Session 16: Properties of materials

Coll
Secure

The properties of a material depend on how the atoms are bonded, the size of the molecules and the arrangement of the atoms and ions. Small molecules have weak forces between them so they have low boiling points and are normally gases. Metals are good conductors of electricity because of their sea of delocalised electrons.

C1

Small molecules

Collins
Secure Science

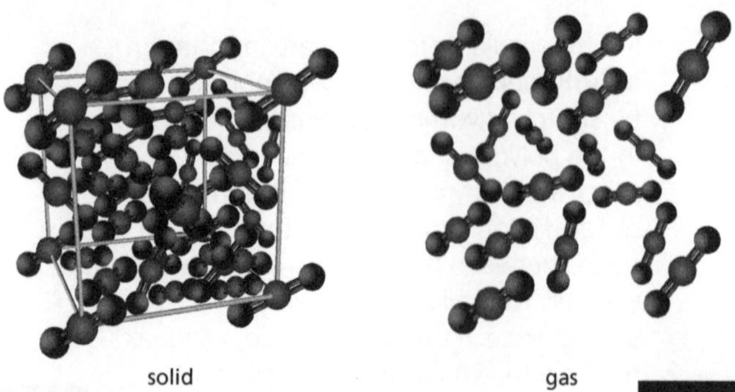

C16.2

Small molecules

Coll
Secure

Substances with small covalent molecules have low melting and boiling points, so usually exist as liquids or gases at room temperature. Solids are formed when cooled to lower temperatures, e.g. carbon dioxide. The weak intermolecular forces between the molecules mean they are easy to separate. They do not conduct electricity as there are no delocalised electrons or ions available to move.

C1

Ionic compounds

Collins
Secure Science

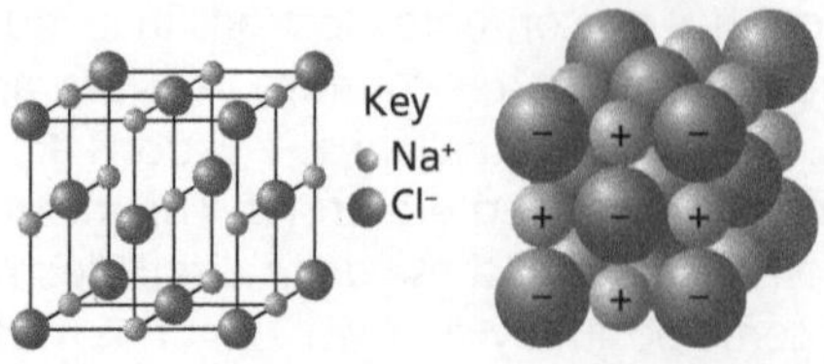

C16.3

Ionic compounds

Col
Secure

Ionic compounds such as sodium chloride form giant crystal lattices. We can use different types of 3D models to represent the structures, such as a ball and stick model or a close packed model.

These models are useful but they have limitations. The ball and stick model is limited because the attractive force between the ions is shown as a solid stick. The close packed model is limited as it shows the ions as solid spheres, when in reality a lot of the ion is empty space.

C1

Ionic compounds

(a) dot and cross diagram

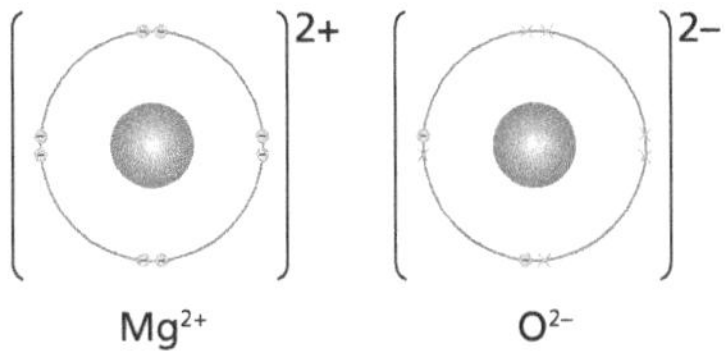

(b) 3D close packed diagram

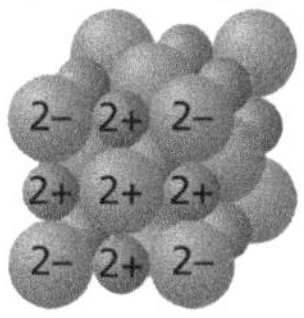

C16.4

Ionic compounds

Collins
Secure Science

Ionic compounds have high melting and boiling points because the forces of attraction holding the ions together are very strong. They conduct electricity when melted or dissolved in water because the ions are free to move and so charge can flow.

The melting point of magnesium oxide, MgO, is 2852 °C. This is explained by its ionic bonding.

Remember that it takes a lot of energy to break the strong forces of attraction between the 2+ and 2– ions.

C16.4

Metals and alloys

Collins
Secure Science

C16.5

Metals and alloys

Collins
Secure Science

The delocalised electrons in metals make them good conductors of electricity. The layers of atoms can easily slide over each other which means they can be bent and shaped. They are **malleable** (can be hammered into shape) and **ductile** (can be pulled into wires). They have high melting and boiling points due to the strong metallic bond.

An **alloy** is a mixture. The different sized atoms twist or distort the layers, making it more difficult for them to slide over each other. This means that alloys are harder than pure metals. For example, steel is an alloy made from iron mixed with carbon. It is used in the construction industry because it is very strong.

Metal	Lead	Tin	Solder
Melting point in °C	327	232	183

Solder is an alloy made from lead and tin. It can be used to join two metals together, due to its low melting point.

C16.5

Giant covalent structures

Collins
Secure Science

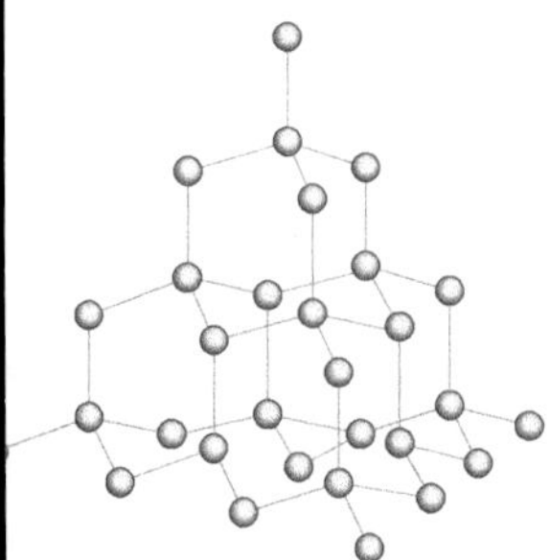

Diamond

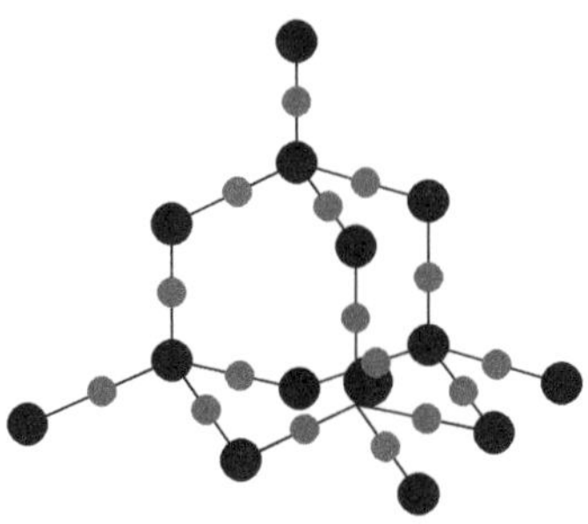

Silicon dioxide

C16.6

Giant covalent structures

Collins
Secure Science

Substances with **Giant covalent molecules** such as diamond (C) and sand (SiO_2) have high melting and boiling points. It takes a lot of energy to break the strong covalent bonds formed between each atom. These materials are very hard and do not conduct electricity as there are no delocalised electrons or ions available.

C16.6

Giant covalent structures

Collins
Secure Science

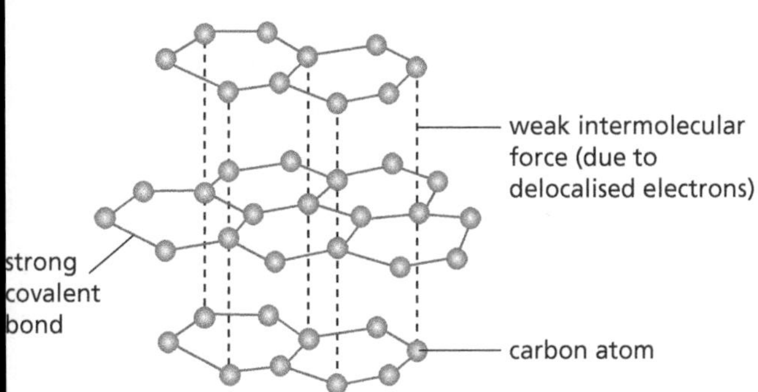

C16.7

Giant covalent structures

Collins
Secure Science

Graphite is also a giant covalent molecule but it is soft and does conduct electricity. It does not follow the rule. This is because, in graphite, carbon atoms form hexagonal rings of carbon atoms in layers. It also has free or delocalised electrons that can move through the structure. It is the free electrons that allow graphite to conduct electricity.

C16.7

Session 17: Elements, compounds and mixtures

Collins
Secure Science

Elements, compounds and mixtures

C17.1

Session 17: Elements, compounds and mixtures

Coll
Secure S

Elements are substances made out of only one type of atom. **Mixtures** are made out of two or more substances that are not chemically bonded together. **Compounds** are substances that contain at least two different elements chemically bonded together. The parts of a compound cannot be easily separated. The parts of a mixture can be separated using the different properties of each part of the mixture. For example, the different colours of ink in a pen can be separated because they have different solubilities.

C17

Elements, compounds and mixtures

Collins
Secure Science

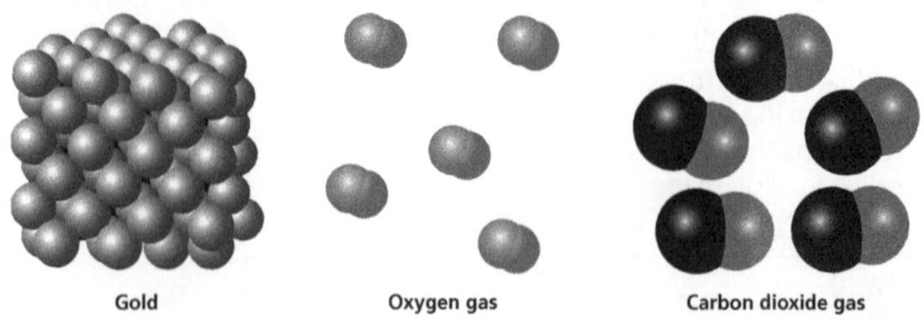

C17.2

Elements, compounds and mixtures

Coll
Secure S

Gold is an element. It only contains atoms of gold. Oxygen gas is an element. It only contains oxygen atoms in its molecules. Carbon monoxide is a compound. Its molecules consist of one atom of carbon chemically bonded to one atom of oxygen. Air is a mixture of oxygen, nitrogen, water vapour and other gases. The chemical properties of each gas in the air remain unchanged. We can separate the gases in the air using **distillation**.

C17

Chromatography

Collins
Secure Science

Carrying out chromatography

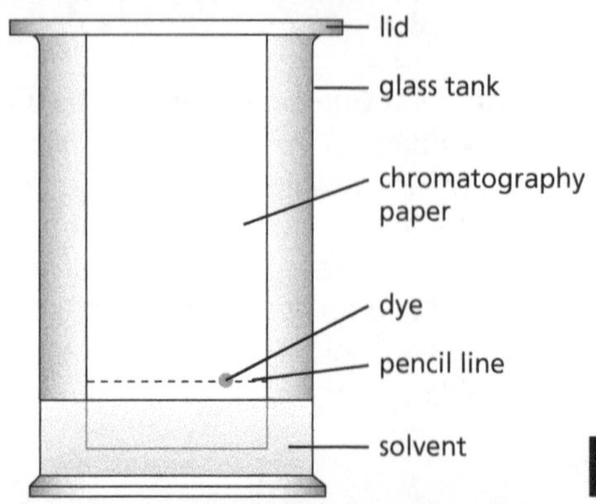

C17.3

Chromatography

Collins
Secure Science

Chromatogram of dyes of food colours

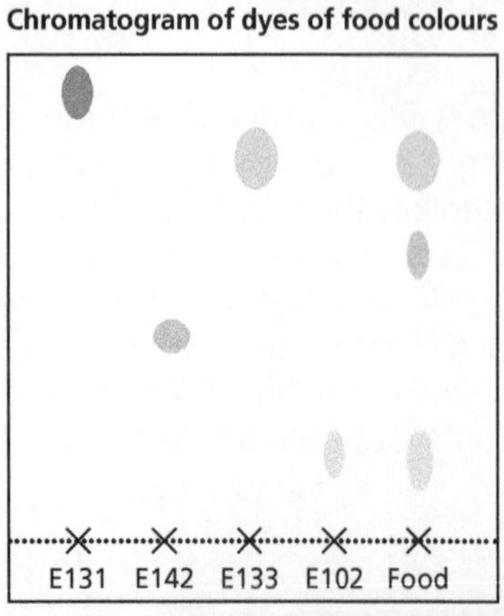

C17.4

Chromatography

Collins
Secure Science

Calculating R_f values

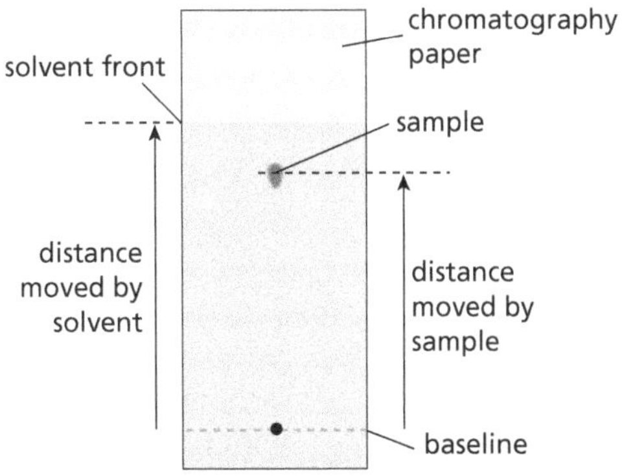

C17.5

Chromatography

Collins
Secure Science

Chromatography uses the property of solubility to separate mixtures of inks, dyes, food colouring or amino acids. During the process, a **solvent** moves through the paper carrying different compounds over different distances. The distance a compound moves depends on how soluble it is in the solvent. The end result is a **chromatogram**, which can be used to identify different parts of the mixture. This is done by calculating and comparing R_f values.

$$R_f = \frac{\text{distance moved by substance}}{\text{distance moved by solvent}}$$

C17.5

Pure and impure substances

Collins
Secure Science

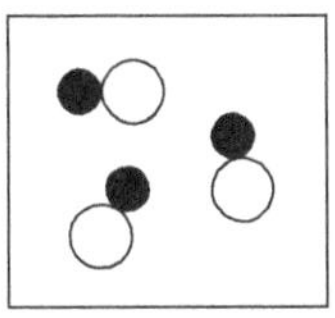

A pure compound

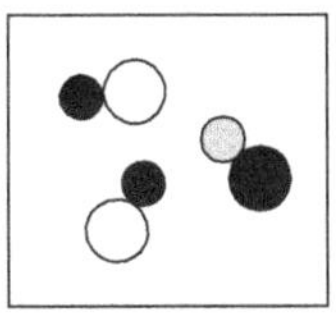

A mixture of compounds

C17.6

Pure and impure substances

Collins
Secure Science

Melting point and boiling point data can be used to tell apart pure substances from mixtures. If a substance is impure the melting point will be *lower* than for the pure substance and the substance will melt over a *broad* range of temperatures. If a substance is *pure* it will melt more *sharply* at a specific temperature. If a substance is *impure* the boiling point will be *higher* than the pure substance.

C17.6

Separation technique

Collins
Secure Science

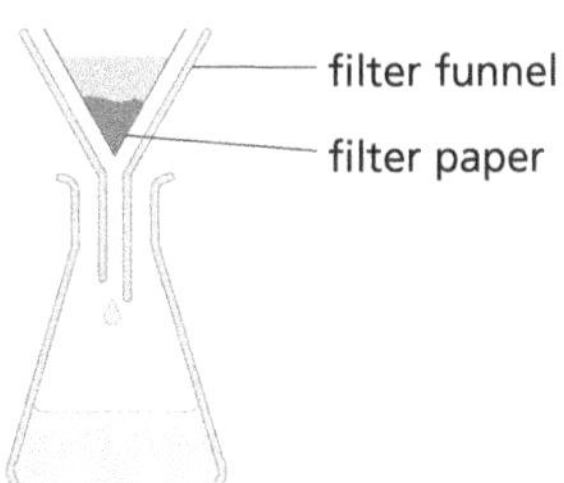

filtration

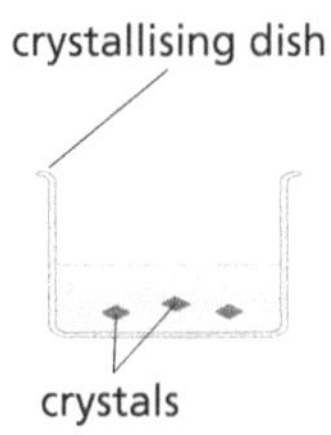

crystallisation

C17.7

Separation technique

Collins
Secure Science

Mixtures can be separated by physical processes that do not involve chemical reactions. For example, we can use a magnet to separate iron, cobalt or nickel from a pile of scrap metals such as copper and aluminium.

C17.7

Separation technique

Collins
Secure Science

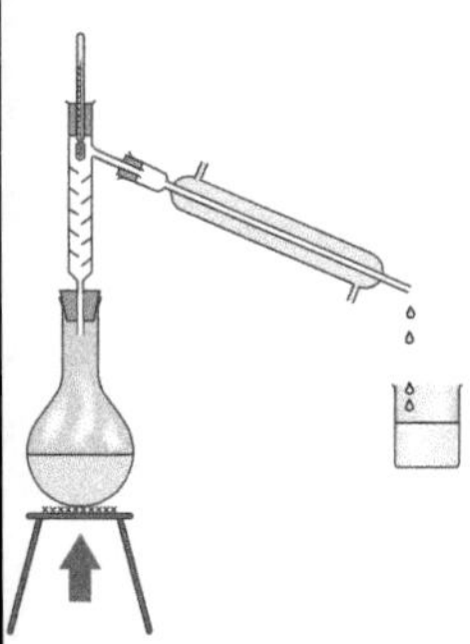

distillation

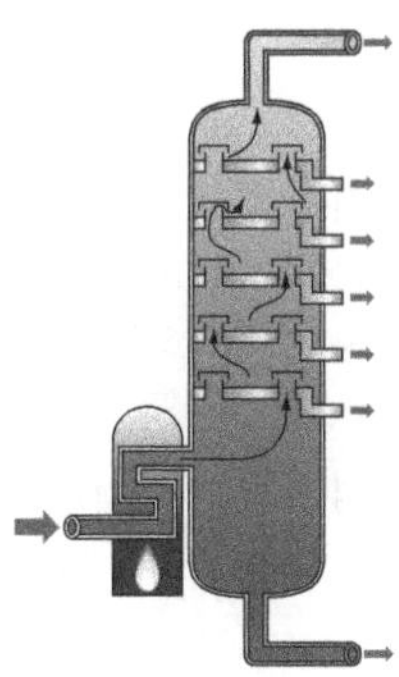

fractional distillation

C17.8

Separation technique

Collins
Secure Science

Filtration is used to separate **insoluble** substances like sand and mud from a **soluble** substance such as sodium chloride. Put the mixture in a beaker with some water. Then pour it into the filter. The insoluble part (sand) will collect on the filter paper. The soluble part (salt solution) will collect in the flask. **Crystallisation** is used to remove water from a solution, leaving solid **crystals** behind. This is done by gently heating the solution so the water evaporates. Distillation is the process of evaporation, followed by condensation. It is used to separate mixtures of liquids with different boiling points, for example alcohol and water. **Fractional distillation** is used to separate **crude oil**. The mixture of liquids is vaporised and compounds with different boiling points condense at different temperatures.

C17.8

Earth chemistry

Earth's atmosphere was formed from chemical reactions and biological processes taking place over millions of years. Crude oil is a fossil fuel made from the remains of organisms that lived millions of years ago. It is a mixture of **hydrocarbons** that can be made into useful substances via physical and chemical processes. The hydrocarbons found in crude oil can be separated into fractions in a process called fractional distillation. Chemical reactions are used to make substances such as **polymers**, plastics and detergents, all with different properties.

Hydrocarbons

Name	Number of carbon atoms	Formula	Displayed formula
Methane	1	CH_4	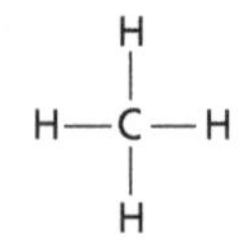
Ethane	2	C_2H_6	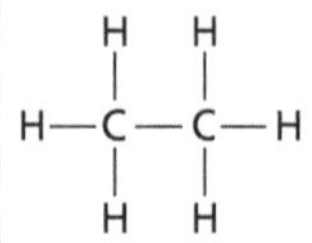

Hydrocarbons

Crude oil is a mixture of hydrocarbon molecules. A hydrocarbon contains only carbon and hydrogen atoms. Most hydrocarbons in crude oil are **alkanes**. Alkanes have the general formula C_nH_{2n+2} where *n* is the number of carbon atoms. The first four alkanes are methane, ethane, propane and butane.

When hydrocarbons burn in oxygen, carbon dioxide and water are produced. Energy is given out to the surroundings in a **combustion** reaction. Fuels extracted from crude oil include petr diesel and methane gas which are used for heating and transport

The equation for the combustion of methane is

methane + oxygen → carbon dioxide + water

$$CH_4 + 2O_2 \rightarrow CO_2 + 2H_2O$$

Earth's atmosphere

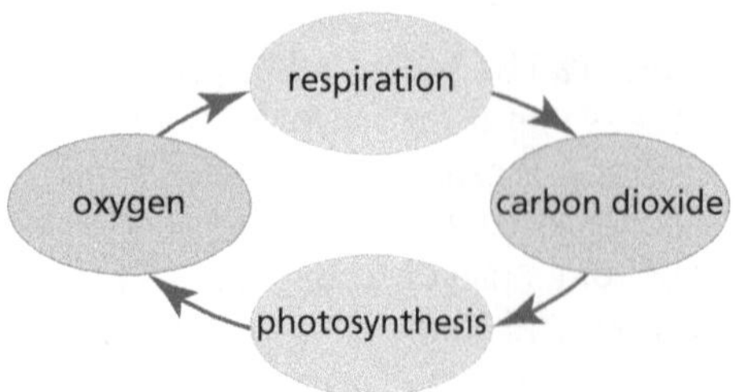

Earth's atmosphere

Air is a mixture of different gases. The early atmosphere contained gases such as ammonia and carbon dioxide from volcanoes. Water vapour condensed to form oceans. Algae used carbon dioxide for photosynthesis and started to produce oxygen. Carbon dioxide dissolved in the oceans.

Earth's atmosphere

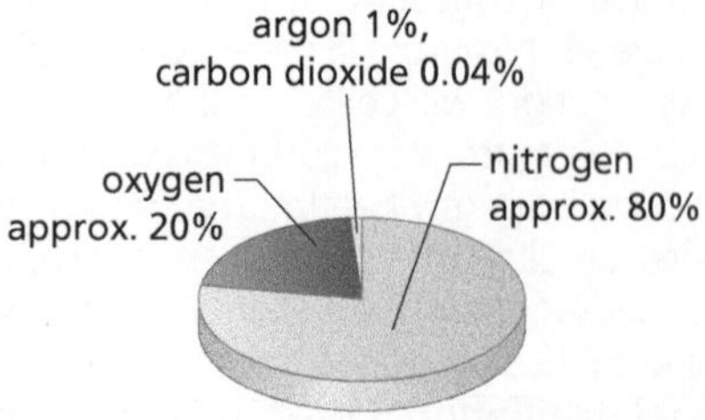

Earth's atmosphere

The proportions of gases in today's atmosphere remain almost constant, apart from water vapour. Plants photosynthesise, releasing oxygen.

carbon dioxide + water → glucose + oxygen

Animals and plants respire, releasing carbon dioxide.

glucose + oxygen → carbon dioxide + water

Polymers

Collins
Secure Science

$C_{16}H_{34} \xrightarrow{\text{cracking}} C_8H_{18} + 2C_3H_6 + C_2H_4$

ng-chain alkane | shorter-chain alkane | alkene | alkene

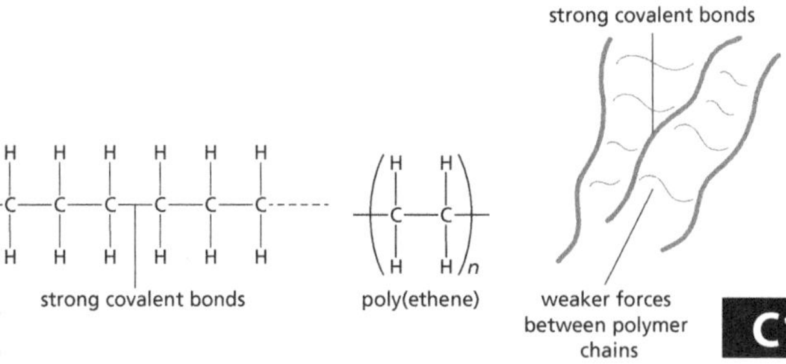

C18.5

Polymers

Collins
Secure Science

Polymers are very large molecules, formed from many smaller molecules called **monomers** in a process called **polymerisation**. Many polymers such as **poly(ethene)** are made from **alkenes**, another hydrocarbon. Alkenes are produced by '**cracking**' some of the larger less useful alkanes produced during fractional distillation of crude oil.

The atoms in polymer molecules are linked to other atoms by strong covalent bonds. Weaker intermolecular forces between the polymer chains hold the chains together. Polymers are usually stretchy solids as the chains can slide over each other when a force is applied.

C18.5

Useful materials

Collins
Secure Science

It is not just fuels that come from crude oil. Different fractions are processed to make plastics, detergents, solvents and bitumen for roads and roofing. Crude oil is used as a raw material in many different industries.

C18.6

Fractional distillation

Collins
Secure Science

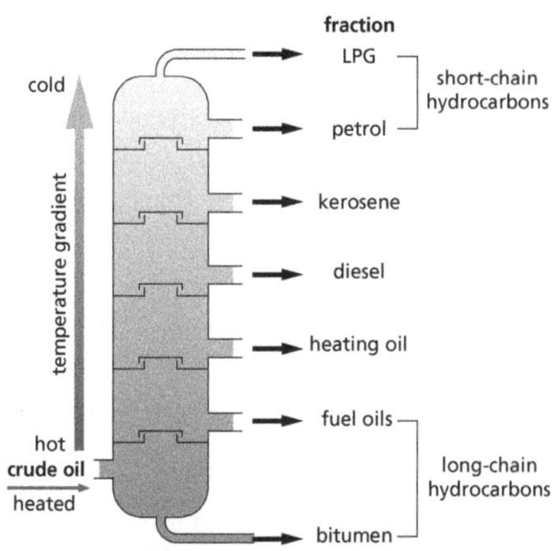

C18.7

Fractional distillation

Collins
Secure Science

Fractional distillation is used to separate crude oil. Crude oil is heated until it forms a **vapour**. The vapour rises up the tower. The different hydrocarbons **condense** at the place in the column where the temperature is just below their **boiling point**. Short-chain hydrocarbons have lower boiling points than long-chain hydrocarbons so rise further up the tower. The different fractions are then collected. A fraction is a group of hydrocarbon molecules with similar boiling points.

C18.7

Session 19: Electrolysis

Collins
Secure Science

Electrolysis

C19.1

Session 19: Electrolysis

Collins
Secure Science

During a chemical reaction, atoms are rearranged. All the atoms that were there at the beginning of the reaction still exist in the products. Chemical changes can be communicated using word and formula equations. Ionic compounds can be rearranged by **electrolysis**. In electrolysis positive ions are attracted to the negative **cathode** and negative ions are attracted to the positive **anode**. This is a useful process for extracting pure reactive metals such as aluminium from its **ore**.

C19.1

Extracting metals

Collins
Secure Science

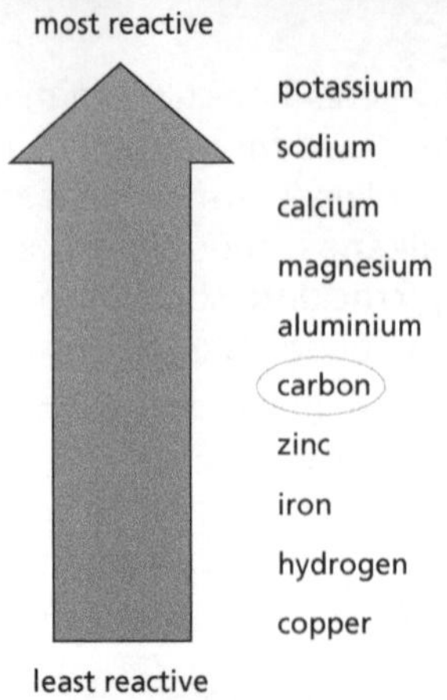

C19.2

Extracting metals

Coll
Secure S

Unreactive metals like gold are found in the earth. Most metals need to be extracted from their ores by chemical reactions. The method of extraction depends on the reactivi

Metals that are less reactive than carbon are extracted using carbon. The metal oxide is **reduced** as carbon removes the oxygen.

zinc oxide + carbon → zinc + carbon dioxide

Metals more reactive than carbon are extracted by electrolys Large amounts of energy are used in the process to melt the compounds and to produce the electricity needed for electrolysis. This means metals such as aluminium are expensive.

C1

Conservation of mass

Collins
Secure Science

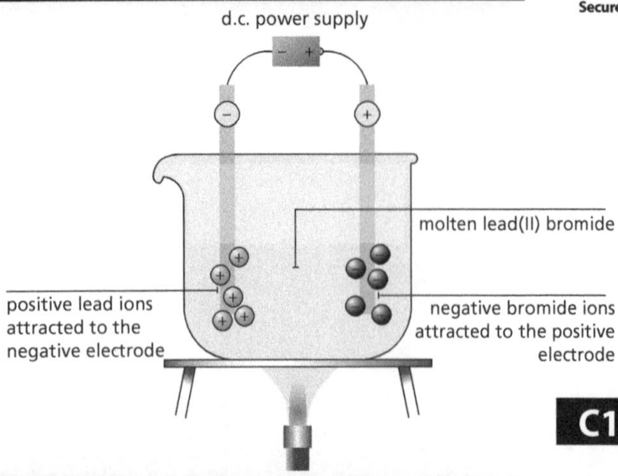

C19.3

Conservation of mass

Coll
Secure S

Solid lead bromide ($PbBr_2$) is a crystal lattice made from Pb^{2+} and Br^- ions. It does not conduct electricity because the ions cannot move. When solid $PbBr_2$ is heated to its melting poin the strong electrostatic forces between the Pb^{2+} and Br^- ions are broken. $PbBr_2$ becomes a liquid. The ions can now move When a direct electric current is passed through the liquid, t ions will move towards the electrodes as it conducts electric

Pb^{2+} ions go to the negative electrode and lead (Pb) is forme

Br^- ions go to the positive electrode, and bromine (Br_2) is formed.

No atoms are lost during electrolysis.
Mass is conserved.

C1

Electrolysis of aqueous solution

Collins
Secure Science

The electrolysis of aqueous solutions is more complicated. As well as ions from the salt there are also hydrogen ions (H^+) and **hydroxide ions** (OH^-) present from the water. This means two ions are attracted to each electrode but only one ion is discharged. The most reactive ion stays in solution and the least reactive ion is discharged. The table summarises which ions are discharged.

For example, in a solution of copper sulfate the ions are: Cu^{2+}, H^+, SO_4^{2-}, OH^-.

Copper is formed at the negative electrode and oxygen is formed at the anode.

In a solution of sodium chloride the ions present are: Na^+, H^+, Cl^-, OH^-.

Hydrogen is formed at the cathode and chlorine is formed at the anode.

C19.4

Electrolysis of aqueous solution

Coll
Secure S

At the cathode	At the anode
If the metal ion is reactive, then hydrogen is formed	If a sulfate group is present, then oxygen is formed
If the metal ion is unreactive, then the metal is formed	If a halide ion is present then the halogen is formed

For example, in a solution of copper sulfate the ions are: Cu^{2+}, H^+, SO_4^{2-}, OH^-.

Copper is formed at the negative electrode and oxygen is formed at the anode.

In a solution of sodium chloride the ions present are: Na^+, H^+, Cl^-, OH^-.

Hydrogen is formed at the cathode and chlorine is formed at the anode.

C1

Electrolysis

Collins
Secure Science

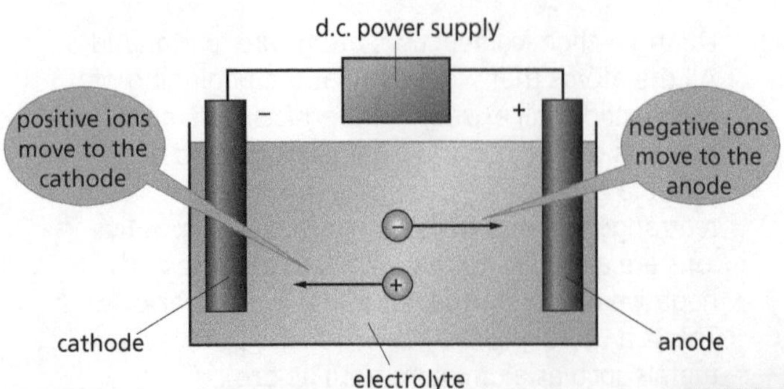

C19.5

Electrolysis

Coll
Secure S

During electrolysis direct current is passed through a molten ionic compound or a solution of an ionic compound, called the **electrolyte**. The ions are attracted towards the electrodes where they are **discharged** (they either gain or lose electrons), producing elements.

C1

Electrolysis

Collins
Secure Science

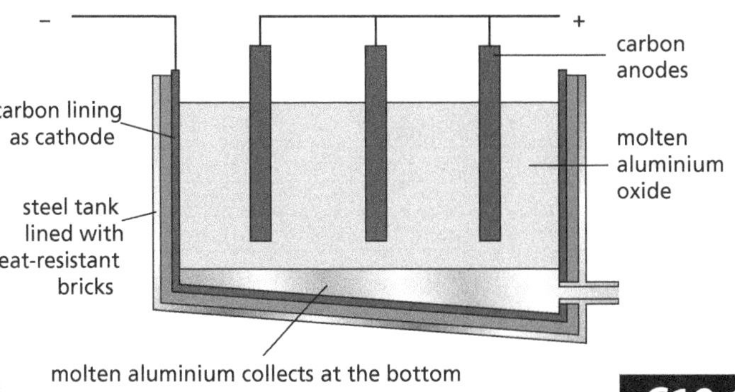

C19.6

Electrolysis

Collins
Secure Science

Aluminium is manufactured by the electrolysis of a molten mixture of aluminium oxide and cryolite using carbon electrodes. The cryolite is added to reduce energy costs by lowering the melting temperature of the aluminium oxide. Aluminium is produced at the cathode (– electrode) and oxygen at the anode (+ electrode). The positive electrode (carbon anode) must be continually replaced because the oxygen reacts with it producing carbon dioxide gas.

C19.6

Session 20: Acids and alkalis

Collins
Secure Science

Acids and alkalis

C20.1

Session 20: Acids and alkalis

Collins
Secure Science

During a chemical reaction, atoms are rearranged. All the atoms that were there at the beginning of the reaction still exist in the products. Chemical changes can be communicated using word and formula equations. When **acids** react with **alkalis**, the hydrogen and hydroxide ions rearrange to make water. The other part of the acid and the metal part of the alkali make a **salt**. The name of the salt is a combination of the name of the acid and the metal part of the alkali.

C20.1

Acids and alkalis

Collins
Secure Science

0–3 = strong acid, 7 = neutral, 12–14 = strong alkali

C20.2

Acids and alkalis

Collins
Secure Science

The **pH** scale is a measure of the acidity or alkalinity of a solution. It can be measured using **universal indicator** or a pH probe. Acids produce hydrogen ions (H^+) in aqueous solutions and have a pH less than 7. Citrus fruit, vinegar and the electrolyte in car batteries are all acidic. Alkaline compounds produce **hydroxide ions** (OH^-) in aqueous solution and have a pH greater than 7. Soap, drain unblocker and oven cleaner are all alkaline.

C20.2

Conservation of mass

Collins
Secure Science

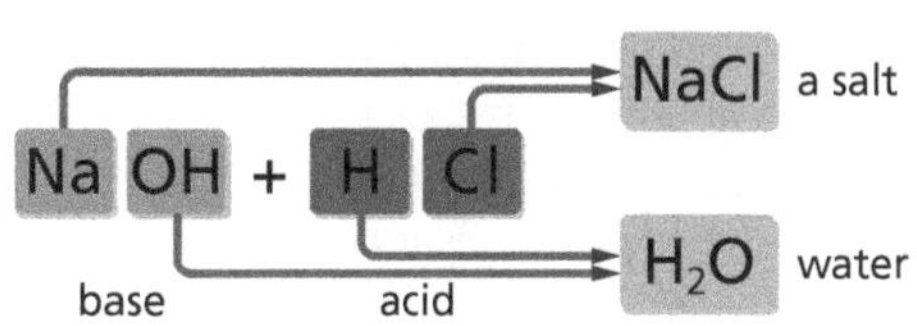

C20.3

Conservation of mass

Collins
Secure Science

During a chemical reaction, all the atoms are rearranged. This is always true for reactions between acids and alkalis even when a gas is made. Remember, gases have mass. If it looks like the mass has gone down during an experiment, it will be because a gas has escaped into the surroundings.

Count the atoms on each side of the equation. They will always be the same.

sulfuric acid + sodium → sodium sulfate + hydrogen

$H_2SO_4(aq) + 2Na(s) \rightarrow Na_2SO_4(aq) + H_2(g)$

nitric acid + calcium carbonate → calcium nitrate + water + carbon dioxide

$2HNO_3(aq) + CaCO_3(s) \rightarrow Ca(NO_3)_2(aq) + H_2O(l) + CO_2(g)$

C20.3

Salts

Making a salt with excess metal oxide or carbonate

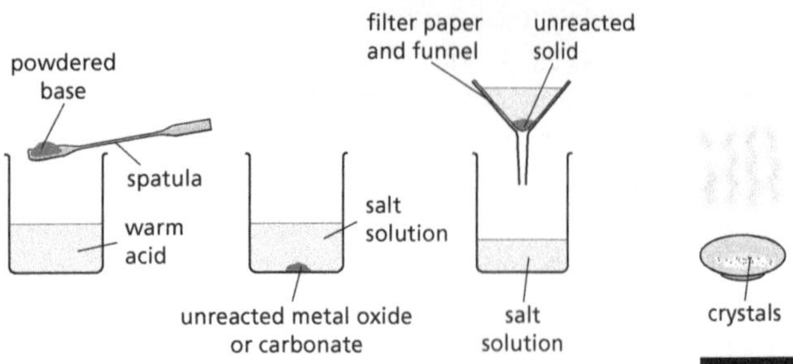

C20.4

Salts

Salt crystals are made when an acid is mixed with excess me
base or carbonate. A base is a metal oxide or alkali. Unreacte
solid is removed by filtration and the salt is left to crystallise.

Ions from alkalis and carbonates		Ions from bases and carbonates		Ions from acids	
Sodium	Na^+	Magnesium	Mg^{2+}	Chloride	Cl^-
Potassium	K^+	Zinc	Zn^{2+}	Nitrate	NO^{3-}
Calcium	Ca^{2+}	Copper	Cu^{2+}	Sulfate	SO_4^{2-}

You can work out the name and formula of a salt from the ions present in the acid and base. For example, zinc oxide an sulfuric acid make the salt zinc sulfate. The formula is $ZnSO_4$. The 2+ charge from the Zn^{2+} ion is cancelled out by the 2– charge from the SO_4^{2-} ion.

C2(

Neutralisation

Acid	Formula	Alkali	Formula
Hydrochloric acid	HCl	Sodium hydroxide	$NaOH$
Nitric acid	HNO_3	Potassium hydroxide	KOH
Sulfuric acid	H_2SO_4	Ammonium hydroxide	NH_4OH
Ethanoic acid	CH_3COOH	Calcium hydroxide	$Ca(OH)_2$

C20.5

Neutralisation

When an acid reacts with a base a **neutralisation** reaction takes place and a salt is produced. A **base** is a metal oxide, hydroxide or carbonate. An **alkali** is a base that is soluble in water and releases OH^- ions.

acid + base → a salt + water

hydrochloric acid + sodium hydroxide → sodium chloride + water

The hydrogen ion from the acid reacts with the hydroxide ion from the alkali to make water. Water is neutral and has a pH of 7.

$$H^+ + OH^- \rightarrow H_2O$$

C2(

General equations for making salts

acid + base → a salt + water

hydrochloric acid + copper oxide → copper chloride + water

acid + metal → a salt + hydrogen

nitric acid + zinc → zinc nitrate + hydrogen

C20.6

General equations for making salts

acid + metal carbonate → a salt + water + carbo dioxid

sulfuric acid + magnesium carbonate → magnesium sulfate + water + carbo dioxid

C2(

Session 21: Rates of reaction

Rates of reaction

C21.1

Session 21: Rates of reaction

A chemical reaction happens when particles collide with enough energy to react. The rate of a reaction is a measure of how fast a product is made or a reactant is used up. The rate of reaction gets faster when the concentration of reactants, the temperature, the surface area of solid reactants or the pressure of reacting gases is increased, or a catalyst is used. A catalyst reduces the amount of energy the particles need to have to react when they collide. A catalyst is not used up in a chemical reaction.

C2

Collision theory

Collins Secure Science

Successful collisions between particles in a reaction

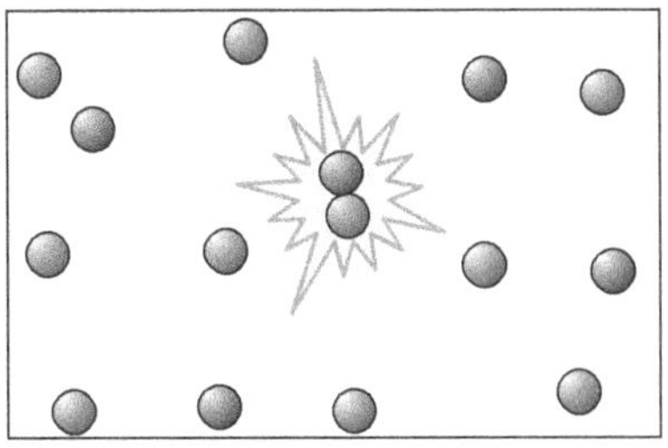

C21.2

Collision theory

Collins Secure Science

Chemical reactions take place when reactant particles successfully collide with each other and form new products. For a collision to be successful particles must collide in the right direction and with enough energy. The minimum amount of energy needed for a successful collision is called the **activation energy**. The rate of reaction is determined by how many successful collisions there are per second. The more successful collisions, the faster the reaction.

C21.2

Taking a closer look at surface area

Collins Secure Science

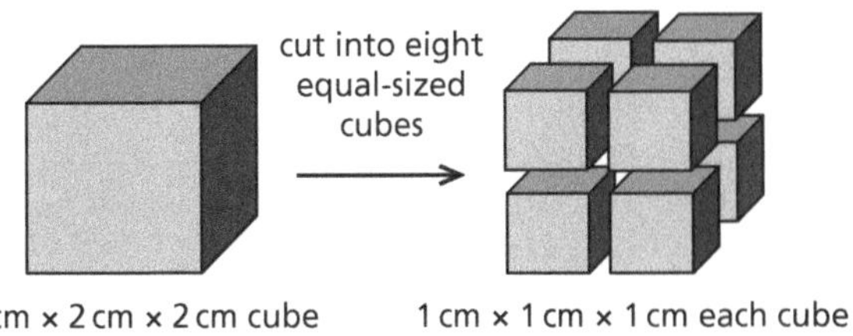

C21.3

Taking a closer look at surface area

Collins Secure Science

Smaller particles of the same total mass have a greater surface area than larger particles.

C21.3

Taking a closer look at surface area

Collins Secure Science

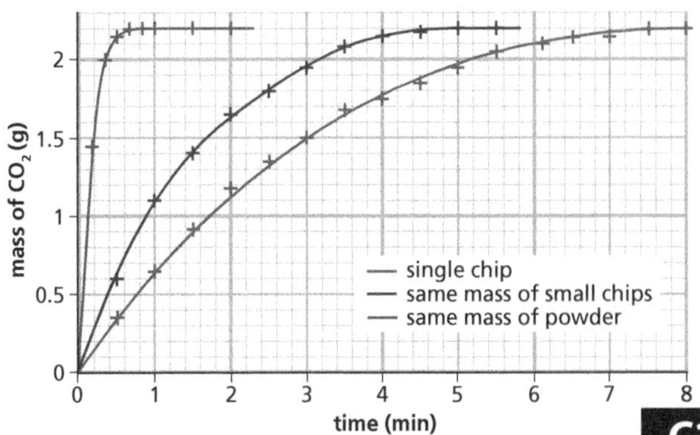

C21.4

Taking a closer look at surface area

Collins Secure Science

This graph clearly shows how the rate of reaction increases as the surface area increases.

C21.4

Reaction profiles

Collins Secure Science

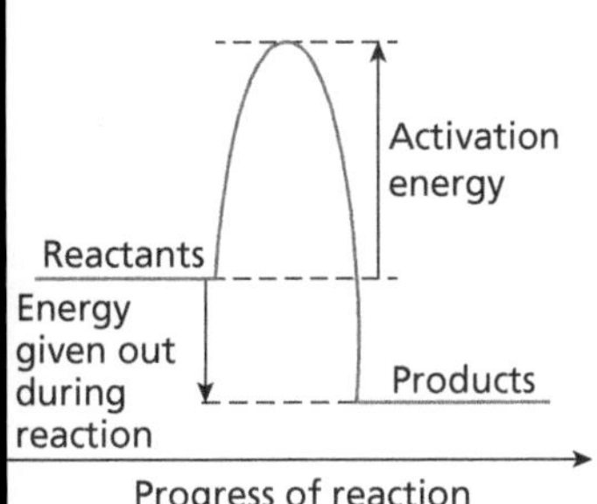

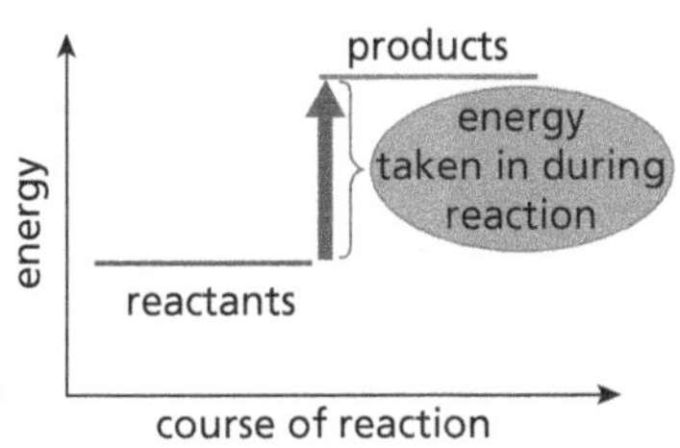

C21.5

Reaction profiles

Collins Secure Science

Reaction profiles show how the energy of the particles changes during the reaction. On the profile above the energy level of the products is lower than the energy level of the reactants. This means energy was given out to the surroundings during the reaction. The reaction is **exothermic**. The profile also shows the activation energy.

When the energy level of the product is higher than that of the reactant, the reaction is **endothermic**. This means that energy has been taken in from the surroundings.

C21.5

Reversible reactions

Collins
Secure Science

reactants $\rightleftharpoons$ products

In some chemical reactions, the products can react to make the original products again. These reactions are **reversible**. When white ammonium chloride is heated, two colourless gases are formed. When the gases cool, they reform into white ammonium chloride.

$$\text{ammonium chloride} \underset{\text{cool}}{\overset{\text{heat}}{\rightleftharpoons}} \text{ammonia + hydrogen chloride}$$

If the rate of the forward reaction is equal to the rate of the backward reaction we say that the reaction has reached **equilibrium**.

C21.6

Rates of reaction

Collins
Secure Science

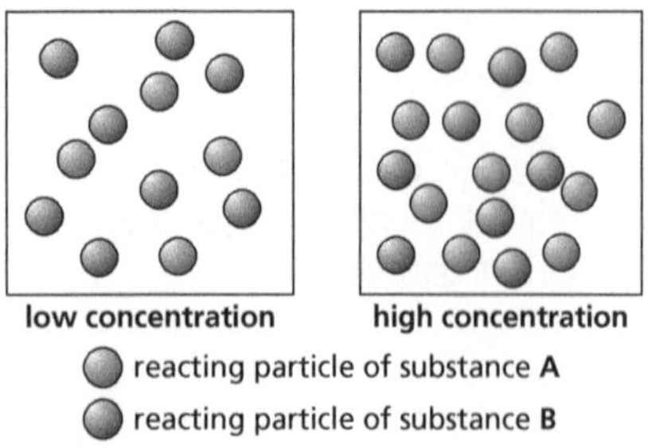

C21.7

Rates of reaction

Col
Secure S

Increasing concentration increases the chances of a successful collision and therefore increases the rate of reaction.

C2

Rates of reaction

Collins
Secure Science

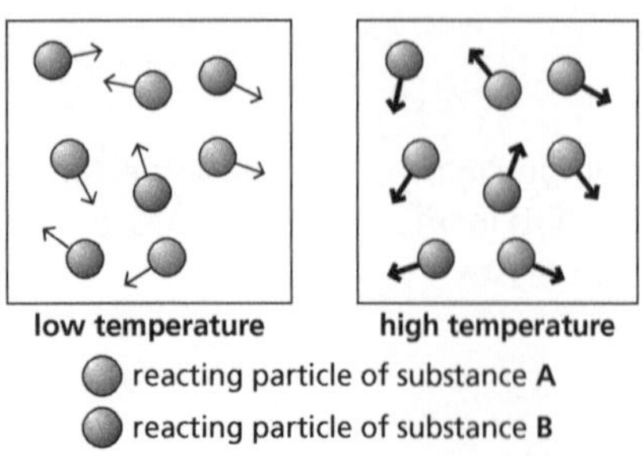

C21.8

Rates of reaction

Col
Secure S

Increasing temperature increases the chances of a successful collision as the particles have more kinetic energy. Increasing temperature therefore increases the rate of reaction.

C2

Rates of reaction

Collins
Secure Science

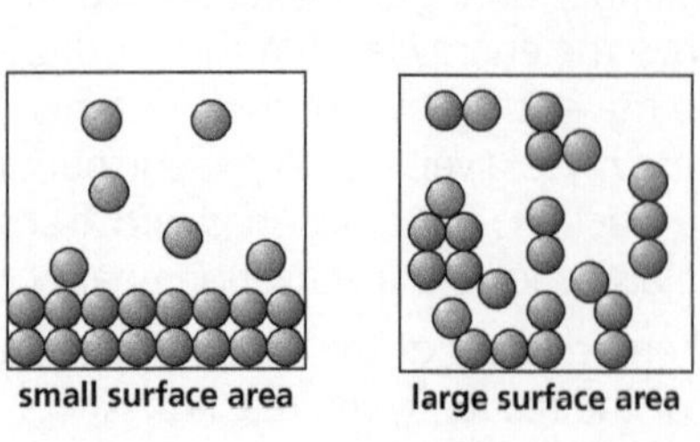

C21.9

Rates of reaction

Col
Secure S

Increasing surface area increases the chances of particles colliding and so also increases the rate of reaction. A 1 g lump of limestone (calcium carbonate) reacts with hydrochloric acid more slowly than 1 g of powered limestone. This is because in the powdered sample, there are more calcium carbonate particles readily available to collide with the hydrochloric acid.

C2

Rates of reaction

Collins
Secure Science

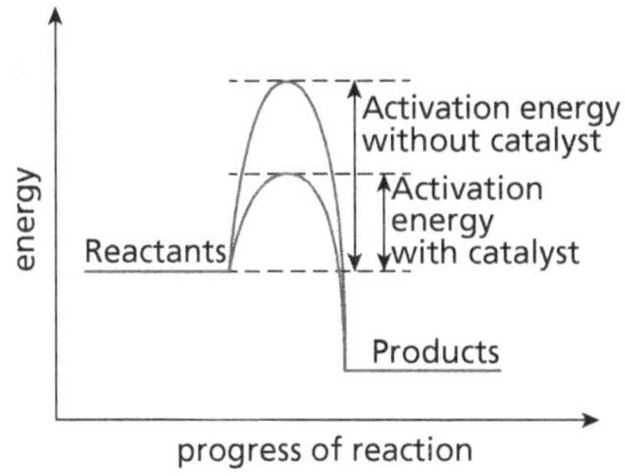

C21.10

Rates of reaction

Collins
Secure Science

Catalysts lower the activation energy of a reaction which increases the chances of a successful collision. Catalysts therefore increase the rate of reaction. Catalysts are not used up themselves in the reaction.

C21.10

Risk assessment

Collins
Secure Science

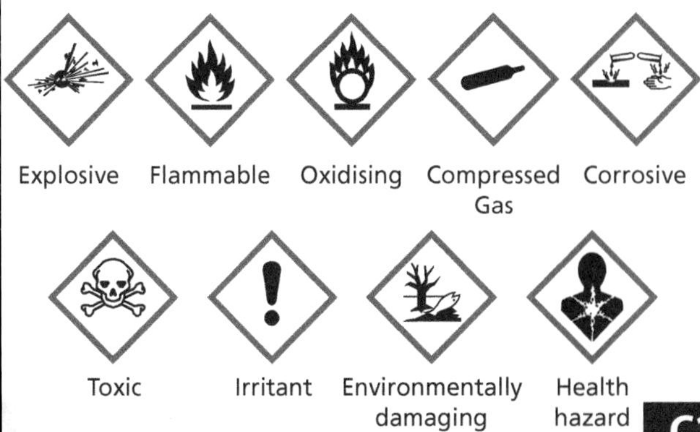

C22.1

Risk assessment

Collins
Secure Science

A hazard is anything which could cause harm: for example, hydrochloric acid or water on the floor. A risk is the chance of harm actually happening. That is, how likely it is that you will get acid in your eye, or slip and fall over as you walk through the pool of water on the floor. When planning an experiment, scientists carry out a risk assessment so that they are aware of all the possible hazards and they think about what they can do to minimise the risk. For example, wearing safety glasses and mopping up the water.

C22.1

nvestigating temperature changes in reacting solutions

Collins
Secure Science

acid + base → a salt + water + energy

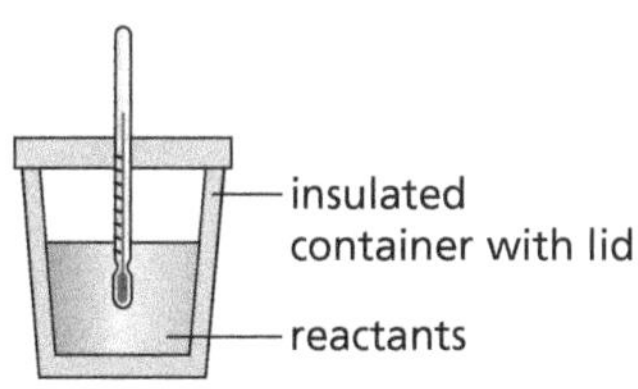

C22.2

Investigating temperature changes in reacting solutions

Collins
Secure Science

The variables in this investigation include the type of base (remember a base is a metal oxide, metal hydroxide or metal carbonate), the type of acid, the concentration of acid or **alkali** (remember an alkali is a soluble base), the size (or surface area) of the pieces of **metal oxide or carbonate**. During the investigation we can only change one variable, the **independent variable**. All others must be kept the same (**control variables**). In this experiment the **dependent variable** is the temperature change. Use a thermometer or temperature sensor to measure the temperature at the start and the end of the reaction.

C22.2

Investigating rates of reaction

Collins
Secure Science

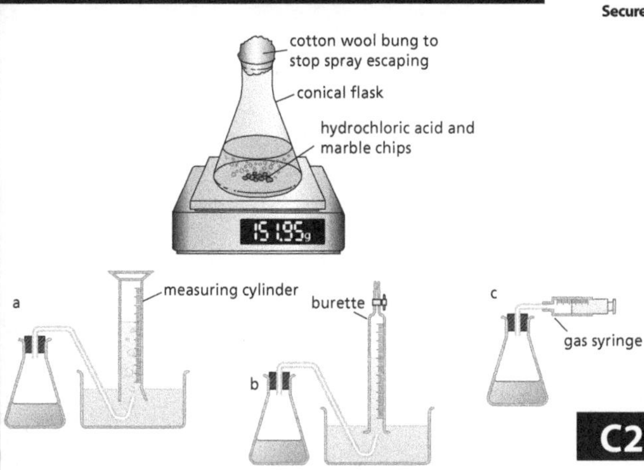

C22.3

Investigating rates of reaction

Collins
Secure Science

The choice of equipment is important when investigating the rate of reaction to get accurate and repeatable results. When hydrochloric acid reacts with marble chips, carbon dioxide gas is produced. We can measure the mass of gas produced over time or the loss in mass of the reactants over time. The mass of gas produced is very small, so a sensitive balance is needed (±0.01 g). There are various ways to collect a volume of gas. Think about the equipment available, how easy it is to set up and the accuracy of the scale.

C22.3

Analysis and purification of water

Collins
Secure Science

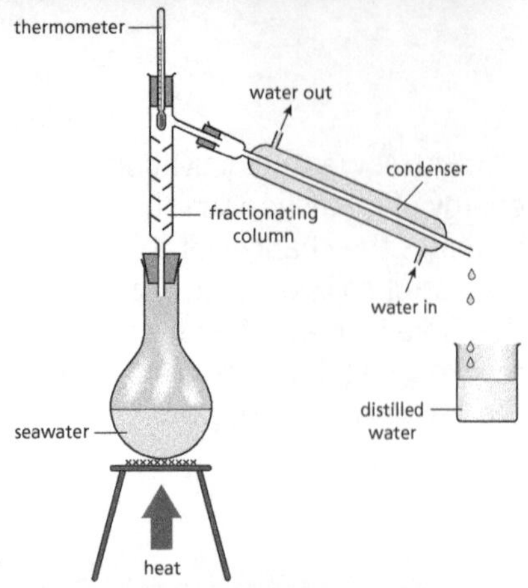

C22.4

Analysis and purification of water

Coll
Secure S

Analytical chemists are like detectives. They have to carry out a series of experiments, process the results and use their knowledge to interpret the results. You need to know which technique to choose. To purify a sample of water, remove any insoluble particles by filtration and then distil most of the filtrate and pour the rest into an evaporating basin and heat it carefully. Any soluble salts will crystallise as the water evaporates.

During distillation, take care when heating, only use a gentle Bunsen flame otherwise dissolved salts in the sample will boil over. Note down the temperature at which the water condenses. It should be 100 °C if it is pure. Test the pH by putting a drop of water onto the universal indicator. It should be 7.

C22

Electrolysis of aqueous solutions

Collins
Secure Science

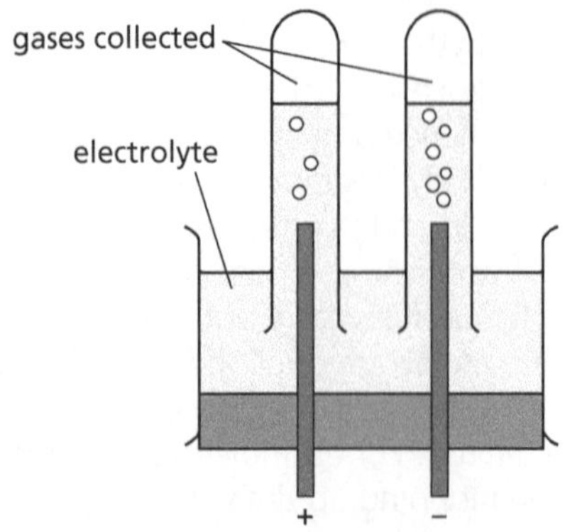

C22.5

Electrolysis of aqueous solutions

Coll
Secure S

The electrolysis of an aqueous solution is often carried out in an electrolysis cell. Using a small test tube you can collect and test the gases formed at the electrodes. The gas is collected by displacement of the electrolyte from the inverted tubes, so you need to make sure the tubes are full of the liquid at the start. Keep the voltage at around 6 V – no higher.

C22

Preparing soluble salts

Collins
Secure Science

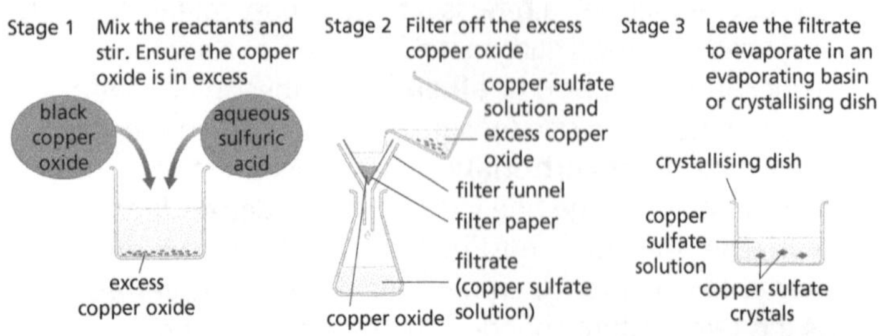

C22.6

Preparing soluble salts

Coll
Secure S

The example shows making copper sulfate crystals. When making a soluble salt you always follow the same method. You will need to identify the correct acid and base or carbonate to make the desired salt. To make a purer product you will need to wash, dry and recrystallise the salt.

C22

Session 23: Forces

Forces

P23.1

Session 23: Forces

A force is a push, pull or twist that causes an object to change speed, change direction or change shape. A force can cause a stationary object to start moving or a moving object to become stationary. When two or more forces act on an object, their effects on the object are added together. If the forces cancel each other out they are said to be balanced, and there is no change in the movement or shape of the object. If the forces are unbalanced, then the movement or shape of the object will change. Newton's First and Second Laws of Motion explain how forces affect a single object. Newton's Third Law of Motion explains that when two objects interact the forces they exert on each other are equal and opposite.

P23.1

Forces and their effects

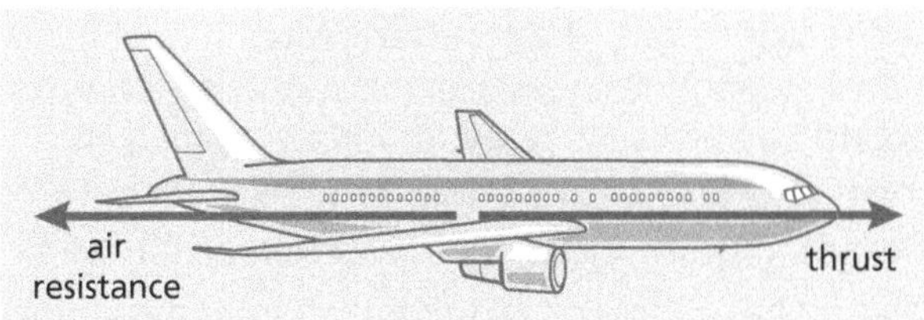

P23.2

Forces and their effects

The thrust of the engines is a push force acting to the right. The **air resistance** between the aircraft and particles of air is a push force acting to the left. The forces are shown as arrows. The two arrows have the same size but act in opposite directions, so they are balanced. This means the aircraft travels at constant speed.

P23.2

Newton's First and Third Laws

P23.3

Newton's First and Third Laws

Newton's First Law explains why the car is travelling at constant speed. The forward force from the engine, is equal and opposite to the combined force of the drag and friction. Newton's First Law states that an object will stay still or keep moving at a constant speed unless it is acted on by an external force, so the resultant of these forces on the car is zero

P23.3

Newton's First and Third Laws

P23.4

Newton's First and Third Laws

When two cars crash, the green car exerts a force F_2 on the blue car. The blue car exerts an equal and opposite force F_1 on the green car. This is an example of **Newton's Third Law** – when two objects interact, they exert forces on each other that are equal in size and opposite in direction.

P23.4

Weight

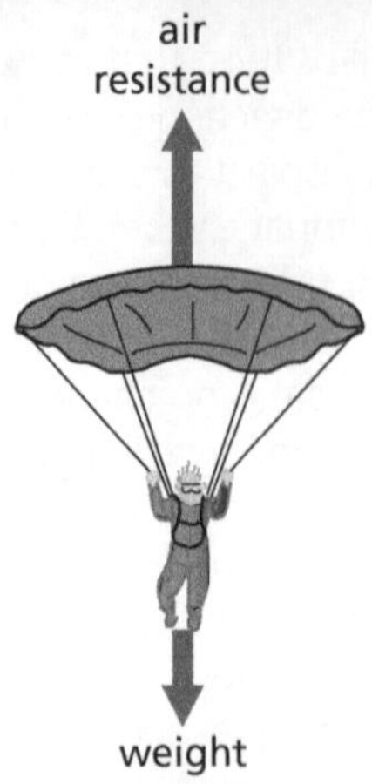

P23.5

Weight

The parachutist is falling downwards because the force of gravity is pulling her towards the Earth. This is called her **weight**. The weight of an object, is given by:

weight (N) = mass (kg) × gravitational field strength (N/kg)

The parachute has just opened and has a large air resistance. At this time, the force of air resistance is greater than the weight of the parachutist. Therefore, there is a resultant force acting upwards and the parachutist is slowing down. However, the force of air resistance depends on the speed of movement, so the upwards force reduces as the parachutist slows down. Eventually the weight of the parachutist and the force due to air resistance will be equal in size and opposite in direction, and the parachutist will fall at a constant speed.

Resultant forces and Newton's Second Law

P23.6

Resultant forces and Newton's Second Law

The diagram shows a driving force of 6000 N to the left. Friction and air resistance combine to produce a force of 2000 N to the right. The **resultant force** is then 4000 N to the left, and the car accelerates to the left. **Newton's Second Law** links the resultant force acting on an object, its mass and its acceleration. The equation is:

resultant force (N) = mass (kg) × acceleration (m/s^2)

If the car has a mass of 2000 kg, then from the equation,
4000 N = 2000 kg × acceleration,

so the acceleration = $\frac{4000\,\text{N}}{2000\,\text{kg}}$ = 2 m/s^2.

Contact and non-contact forces

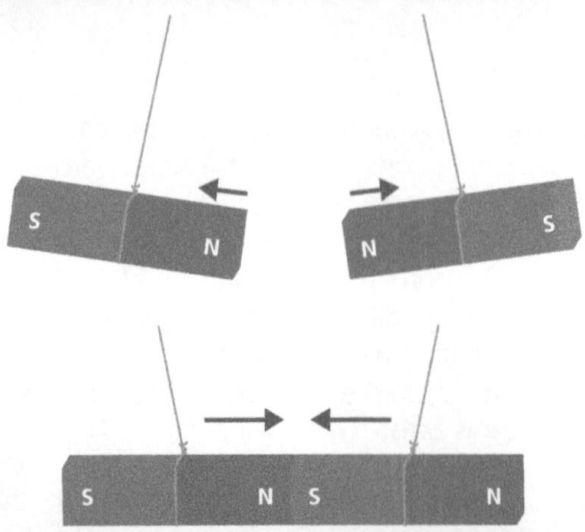

P23.7

Contact and non-contact forces

The air resistance, in the diagram of the aircraft, is an example of a **contact force**. The air particles hit the surface of the aircraft and oppose the motion of the aircraft. **Friction** and **tension** are also contact forces.

Magnetism is an example of a **non-contact force**, as are gravity and electrostatic forces.

Session 24: Motion

Motion

P24.1

Session 24: Motion

Scalar quantities, such as mass, only have size (or **magnitude**). Quantities such as force, have size and direction and are called **vector quantities**. The motion of objects can be described in terms of distances, speeds, velocities and accelerations. The best way to show the motion of an object is by using a **distance–time graph** or a **velocity–time graph**.

Speed, velocity and acceleration

Collins Secure Science

Speed is a measure of how fast an object is moving. Speed is a scalar quantity. The equation for speed is:

$$\text{speed (m/s)} = \frac{\text{distance travelled (m)}}{\text{time (s)}}$$

P24.2

Speed, velocity and acceleration

Collins Secure Science

Velocity is the speed of an object in a given direction and it is a vector quantity. **Acceleration** is the rate at which an object changes velocity. Acceleration is a vector quantity. The equation for acceleration is:

$$\text{acceleration (m/s}^2\text{)} = \frac{\text{change in velocity (m/s)}}{\text{time taken (s)}}$$

P24.2

Stopping distances

Collins Secure Science

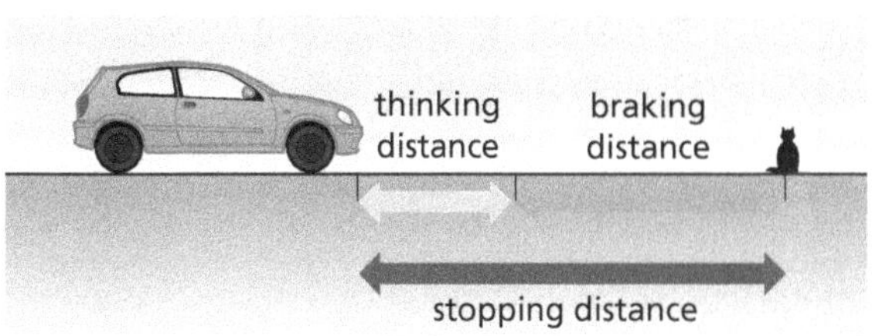

P24.3

Stopping distances

Collins Secure Science

A car's **stopping distance** is the total distance the car travels during the time it takes to stop.

$$\text{stopping distance (m)} = \text{thinking distance (m)} + \text{braking distance (m)}$$

Thinking distance depends on the:

- speed of the car
- reaction time of the driver (affected by alcohol/drugs/tiredness/distractions).

Braking distance depends on the:

- speed of the car
- condition of the brakes/tyres
- road surface (dry/wet/icy/loose/firm).

For a given braking force the greater the speed of the vehicle, the greater the stopping distance.

P24.3

Scalar and vector quantities

Collins Secure Science

Scalar quantities	Vector quantities
Distance	Displacement
Speed	Velocity
Mass	Weight
Time (interval)	Force
Area, Volume	Acceleration
Density	Lift, Drag, Thrust
Energy	Magnetic field
Work	
Power	
Temperature	

P24.4

Terminal velocity

Collins Secure Science

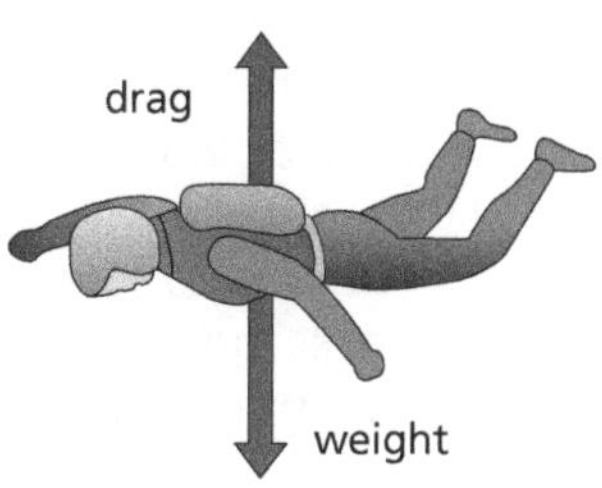

P24.5

Terminal velocity

Collins Secure Science

When a skydiver falls out of a plane, her weight causes the skydiver to accelerate downwards. As the speed of the skydiver increases, so does the air resistance force (the drag). Eventually, the drag force acting upwards, is equal and opposite to the weight acting downwards. The forces are balanced, and the skydiver stops accelerating and travels at a constant speed called the **terminal velocity**.

P24.5

Graphs of motion

Collins
Secure Science

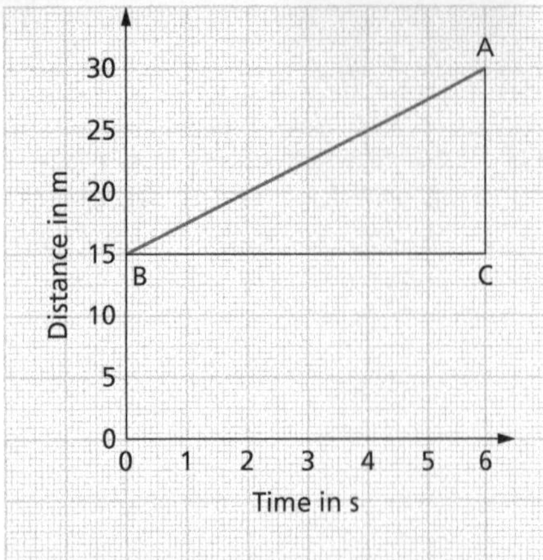

P24.6

Graphs of motion

Coll
Secure S

The distance–time graph shows an object moving at constant speed. At first, at point B, the object is 15 m from an observer. It then moves another 15 m further away from the observer in 6 seconds. The **gradient** (or slope) of a distance–time graph is the speed of the object.

$$\text{The gradient of the line} = \frac{AC}{BC} = \frac{(30\,\text{m} - 15\,\text{m})}{(6\,\text{s} - 0\,\text{s})} = \frac{15\,\text{m}}{6\,\text{s}} = 2.5\,\text{m/s}$$

This means that the speed of the object is 2.5 m/s.

Any **horizontal** line shows that the object is stationary.

P24

Graphs of motion

Collins
Secure Science

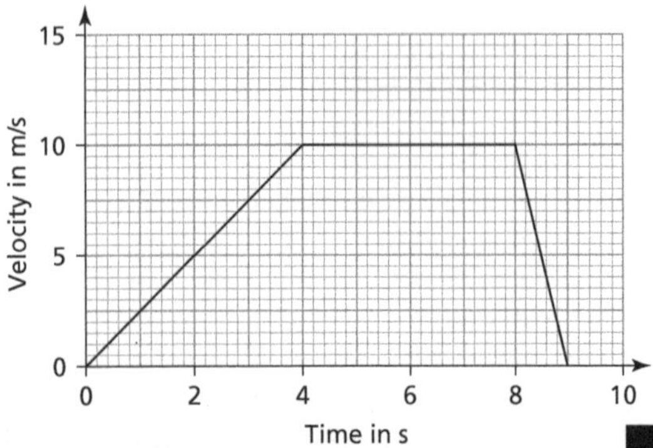

P24.7

Graphs of motion

Coll
Secure S

Acceleration can be calculated from the gradient of a velocity–time graph. This example shows the change in velocity of a remote-controlled car during a race. The acceleration between 0 and 4 seconds can be calculated using the gradient of the line.

$$\text{acceleration (m/s}^2\text{)} = \frac{\text{change in velocity (m/s)}}{\text{time taken (s)}} = \frac{(10 - 0\,\text{m/s})}{4\,\text{s}} = 2.5\,\text{m/s}$$

The horizontal part of the graph shows the car moving at a constant velocity of 10 m/s. The downward slope from 8 to 9 seconds shows the car **decelerating** (slowing down).

P24

Session 25: Waves

Collins
Secure Science

Waves

P25.1

Session 25: Waves

Coll
Secure S

Waves transfer energy. **Longitudinal waves** (e.g. sound waves) cause the particles of the substance the wave is travelling through to move backwards and forwards. **Transverse waves** (e.g. water waves) make the particles move at a right angle to the direction the wave is travelling. Different parts of waves can be measured to understand the properties of waves.

P25

Waves

Collins
Secure Science

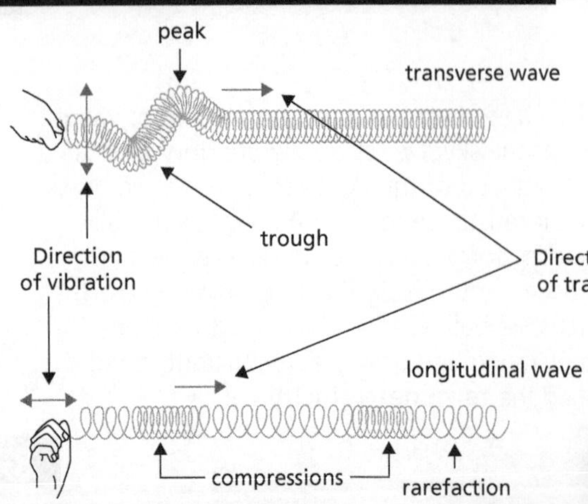

P25.2

Waves

Coll
Secure S

The diagram shows a transverse wave and a longitudinal wave being generated on a slinky. For the transverse wave, the coils move at right angles to the direction that the wave is travelling. This causes peaks and troughs to form. For the longitudinal wave, the coils move backwards and forwards, forming **compressions** and **rarefactions** in the same direction as the direction of travel. All waves transfer energy.

P25

The electromagnetic spectrum

Collins Secure Science

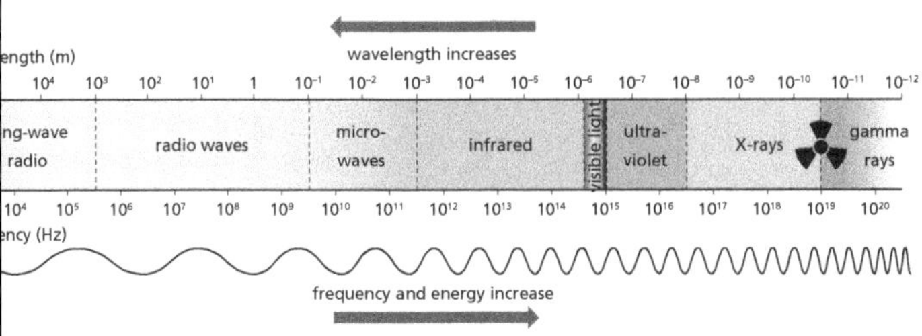

P25.3

The electromagnetic spectrum

Collins Secure Science

The **electromagnetic spectrum** is a family of transverse waves. The waves all travel at the speed of light through a vacuum, and transfer energy from a source to an absorber. They are arranged into seven groups by order of wavelength, frequency and energy. In decreasing order of wavelength (and increasing frequency/energy), they are: **radio waves**, **microwaves**, **infrared**, **visible light**, **ultraviolet**, **X-rays** and **gamma rays**.

P25.3

Properties and uses of the electromagnetic spectrum

Collins Secure Science

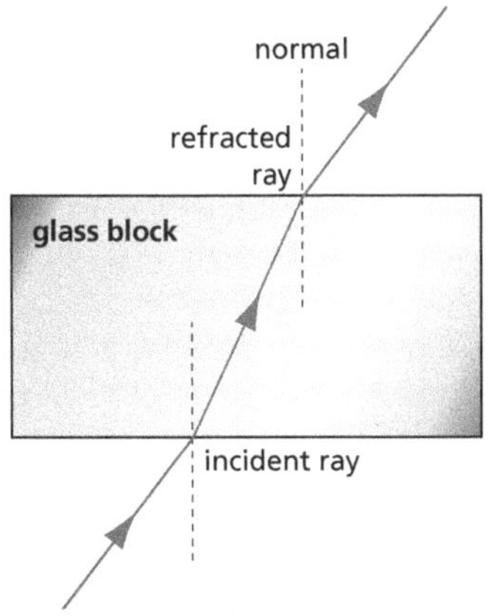

P25.4

Properties and uses of the electromagnetic spectrum

Collins Secure Science

The diagram shows the **refraction** of a beam of light through a glass block. All waves undergo refraction. Refraction allows infrared and visible light to be used in fibre optics. Radio waves, microwaves, infrared and visible light are all used in communications; and microwaves and infrared are both used for cooking. Ultraviolet, X-rays and gamma rays have the highest energies and can be hazardous to the human body, harming or killing cells. X-rays and gamma rays can be used for medical treatments and imaging. The effects of these waves depend on the size of the dose received by the body.

P25.4

Properties of waves

Collins Secure Science

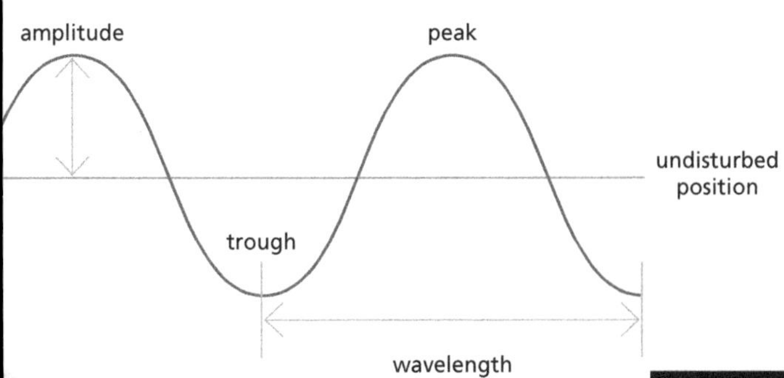

P25.5

Properties of waves

Collins Secure Science

The diagram shows two key properties of a wave. The **wavelength** (in m) is the distance from one point on a wave to the equivalent point on the next wave. The **amplitude** (in m) is the maximum displacement of a point on a wave from its undisturbed position. The **frequency** (in hertz, Hz) of a wave is the number of complete waves passing a point in 1 second. Frequency and wavelength are connected to the wave speed (in m/s) by the wave equation:

wave speed (m/s) = frequency (Hz) × wavelength (m)

P25.5

Session 26: Energy

Collins Secure Science

Energy

P26.1

Session 26: Energy

Collins Secure Science

Energy stores are needed for actions. Energy is transferred from one store to another during an action. A force causes the transfer of energy from one store to another. For example, when a ball is dropped the gravitational potential energy stored in the ball is transferred to kinetic energy. The rate of transfer of energy is called **power**. More powerful devices can transfer more energy per second.

P26.1

Energy stores

Collins
Secure Science

P26.2

Energy stores

Coll
Secure S

Energy can be stored in different forms, including kinetic energy and gravitational potential energy.
The kinetic energy of an object depends on:

- the mass of the object
- and its speed.

The equation for kinetic energy is:

$$\text{kinetic energy (J)} = 0.5 \times \text{mass (kg)} \times \text{speed (m/s)}^2$$

$$E_k = 0.5 \times m\,v^2$$

P26

Energy stores

Collins
Secure Science

P26.3

Energy stores

Coll
Secure

The gravitational potential energy of an object depends on

- the mass of the object
- its height above the ground
- and the **gravitational field strength**.

The equation for gravitational potential energy is:

$$\text{gravitational potential energy (J)} = \text{mass (kg)} \times \text{gravitational field strength (N/kg)} \times \text{height (m)}$$

$$E_p = m\,g\,h$$

P26

Power and efficiency

Collins
Secure Science

Power is the rate of transfer of energy.
The equation for power is:

$$\text{power (W)} = \frac{\text{energy transferred (J)}}{\text{time (s)}}$$

P26.4

Power and efficiency

Coll
Secure S

When energy transfers from one store into other stores, some energy is always wasted, usually through heating. This energy is not lost but is transferred to a less useful energy store.
The **efficiency** of an energy transfer is given by the equation:

$$\text{efficiency} = \frac{\text{useful output energy transfer}}{\text{total input energy transfer}}$$

or, in terms of power:

$$\text{efficiency} = \frac{\text{useful power output}}{\text{total power input}}$$

P26

Energy resources

Collins
Secure Science

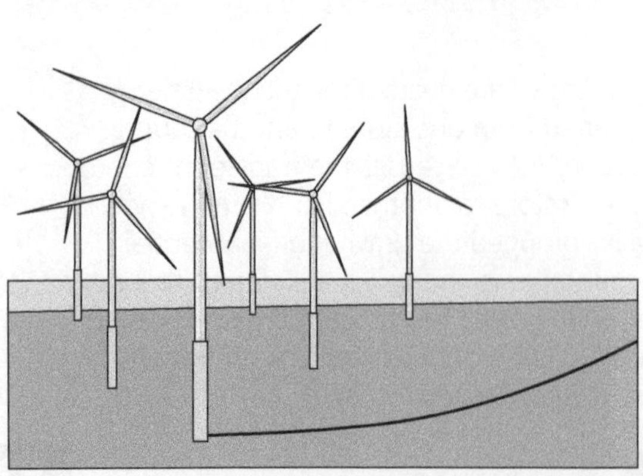

P26.5

Energy resources

Coll
Secure S

Energy resources are generally classified into **non-renewable resources** (coal, oil, gas and nuclear) and **renewable resources** (biofuels, hydroelectricity, geothermal, tidal, solar, wind and wave energies).
Energy resources are used for heating, transport and electricity generation.

When choosing an energy resource for a purpose, the following should be considered:

- the environmental issues associated with the resource
- the reliability of the resource
- any economic considerations.

P2

Energy transfers

Collins Secure Science

Energy can be transferred from one store to other stores in different ways, such as:

- by heating
- electrically
- by radiation
- mechanically.

P26.6

Energy transfers

Collins Secure Science

The gravitational potential energy of a falling object can be mechanically converted into kinetic energy. If all the gravitational potential energy is converted into kinetic energy, then:

$$\tfrac{1}{2} \times \text{mass} \times \text{speed}^2$$
$$= \text{mass} \times \text{gravitational field strength} \times \text{height}$$

P26.6

Session 27: Electricity

Collins Secure Science

Electricity

P27.1

Session 27: Electricity

Collins Secure Science

An electrical circuit is a closed loop of conducting material with at least one **cell** and a component (e.g. a lamp). The cell is the source of electrical energy. Electric **current** is the rate of flow of **charge** through the circuit. As current flows, it is opposed by the structure of the circuit causing **resistance**. In a **series circuit**, the current is the same all the way round the circuit. A **parallel circuit** has different loops for the current to flow through. The total current of all the loops in a parallel circuit is equal to the current flowing through the cell.

P27.1

Electrical circuits

Collins Secure Science

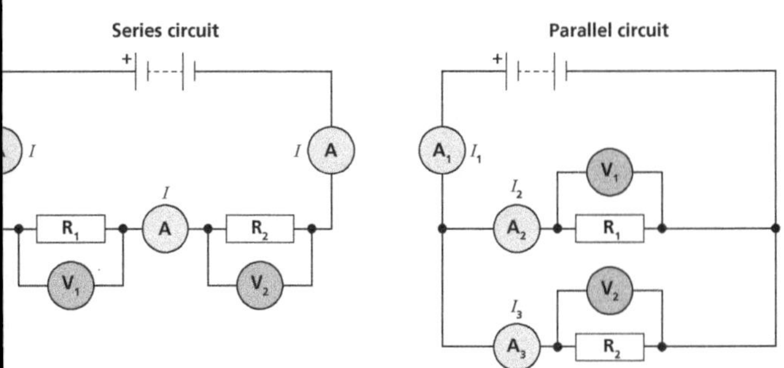

P27.2

Electrical circuits

Collins Secure Science

In the series circuit, components, like **resistors**, are connected in a closed loop. The current, I, is the same through each component. The total **potential difference** (voltage) of the power supply, V, is shared between the components, so $V = V_1 + V_2$. In the parallel circuit, components are connected across each other, using junctions. The potential difference across each component is the same, so $V_1 = V_2$. The total current flowing through the power supply, I_1, is the sum of the currents through each component, or $I_1 = I_2 + I_3$.

P27.2

Resistance of a wire

Collins Secure Science

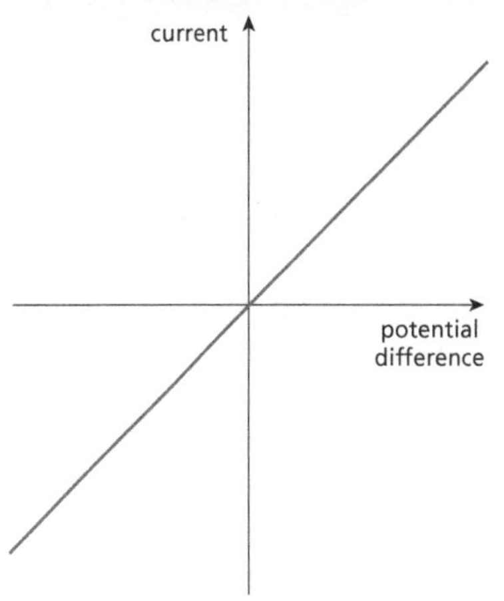

P27.3

Resistance of a wire

Collins Secure Science

An I–V (current–potential difference) graph for a length of wire is shown. The wire obeys the equation $V = IR$, with a constant value of Resistance, R. A wire with a lower resistance would have a steeper gradient (slope). The resistance of a wire depends on the length of the wire: a longer wire has a higher resistance than a shorter similar wire. If the length of the wire is double, then its resistance doubles.

P27.3

Mains electricity

Collins
Secure Science

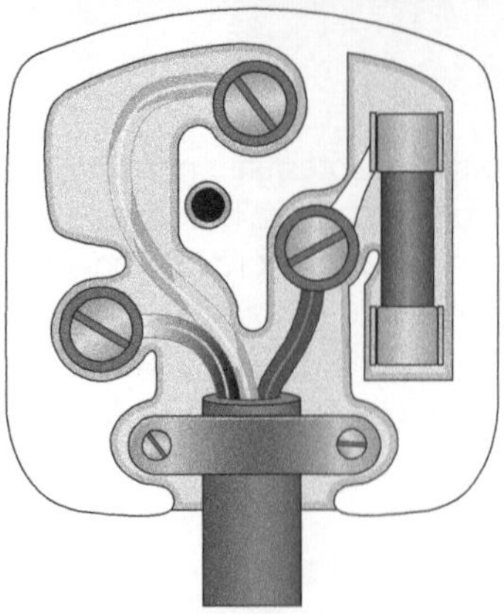

P27.4

Mains electricity

Coll
Secure S

Mains electricity is supplied as an **alternating current** (ac), at a frequency of 50 Hz and a potential difference of 230 V. Mains cables have three coloured wires:

- brown – the live wire (which carries the potential difference).
- blue – the neutral wire (which completes the circuit).
- green/yellow stripes – the earth wire (a safety wire, that stops appliances from becoming live when there is a fault).

P27

Resistance

Collins
Secure Science

The resistance of a component is its opposition to the flow of current through it. Resistance is measured in ohms (Ω). Resistance can be calculated using the equation:

potential difference (V) = current (A) × resistance (Ω)

$$V = I \times R$$

P27.5

Resistance

Coll
Secure S

In the series circuit, the total resistance, R, of the circuit is the sum of resistances, R_1 and R_2, or $R = R_1 + R_2$.

In the parallel circuit, the total resistance is less than the resistance of the lowest individual resistor.

P27

Electrical power

Collins
Secure Science

The power of an electrical device is the rate at which it converts energy supplied in the form of electricity into other more useful forms of energy, such as light, heat or kinetic energy. Power is measured in watts (W).

P27.6

Electrical power

Coll
Secure S

Power is given by the equation:

power (W) = potential difference (V) × current (A)

Power can also be worked out using current and resistance:

power = (current)^2 (A)^2 × resistance (Ω)

P27

Session 28: Electromagnetism

Collins
Secure Science

Electromagnetism

P28.1

Session 28: Electromagnetism

Coll
Secure S

Magnetism is a non-contact force between two poles of a **magnet**. The **magnetic fields** around magnets can be drawn using magnetic field lines. The magnetic field is strongest at the poles of a magnet. **Electromagnets** are formed when an electric current flows through a conductor such as a metal wire. The shape of the magnetic field around an electromagnet depends on the shape of the electromagnet.

P28

Magnetic poles

Collins
Secure Science

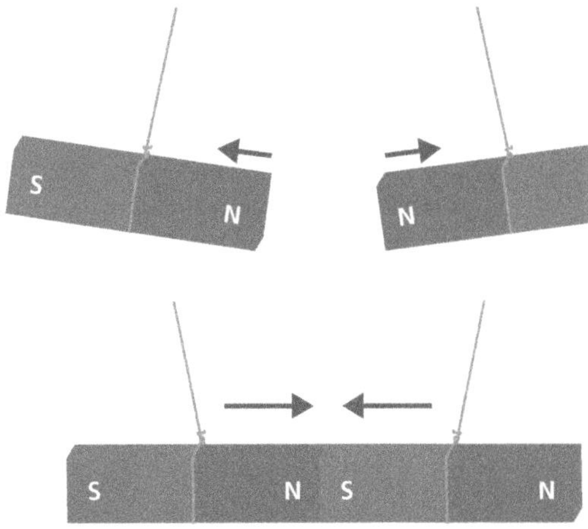

P28.2

Magnetic poles

Collins
Secure Science

Permanent magnets produce their own magnetic field, and they have a north pole and a south pole. When they are brought together, magnetic **poles** exert a force on each other. Like poles **repel** each other and unlike poles **attract** each other. A permanent magnet can make objects that are made of the materials iron or steel, magnetic – this is called induced magnetism. Induced magnetism always causes a force of attraction. If the permanent magnet is moved away from the **induced magnet**, the induced magnet quickly loses its magnetism.

P28.2

Electromagnets

Collins
Secure Science

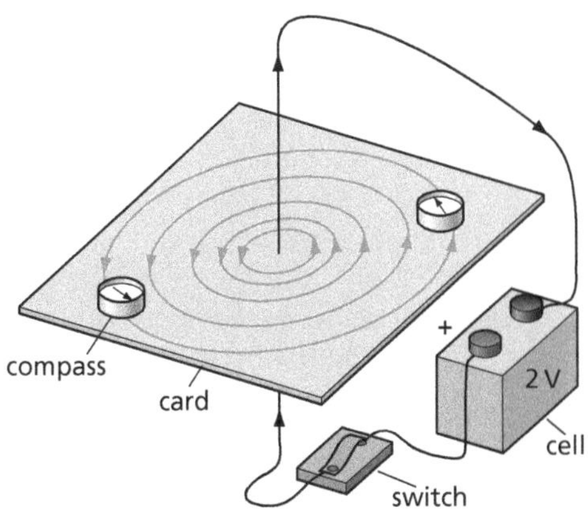

P28.3

Electromagnets

Collins
Secure Science

When an electric current flows through a metal wire, a magnetic field is produced around the wire. The magnetic field lines form **concentric circles** around the wire. Coiling the wire produces an electromagnet. The strength of the magnetic field around an electromagnet depends on the size of the electric current, the number of turns of the coil of wire and the distance away from the electromagnet. A **solenoid** is a long, coiled electromagnet. The magnetic field inside the solenoid is **uniform** and can be increased in strength by adding an iron core.

P28.3

Magnetic fields around electromagnets

Collins
Secure Science

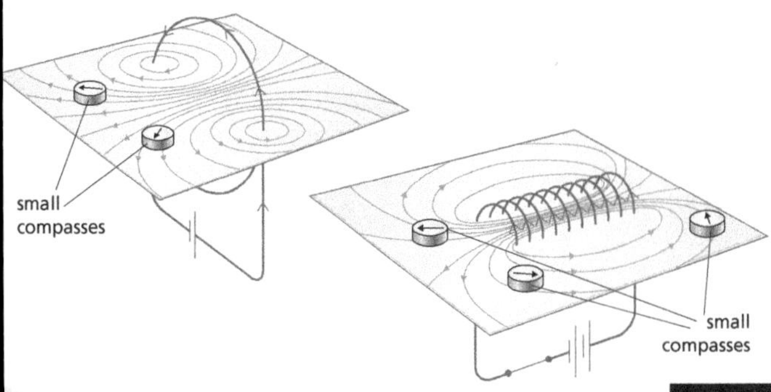

P28.4

Magnetic fields around electromagnets

Collins
Secure Science

The diagram shows the shape of the magnetic fields around a single coil (on the left) and a solenoid (on the right). The magnetic field of the solenoid is like the magnetic field around a permanent magnet. One end of the solenoid acts like a north pole and the other end like a south pole. Electromagnets have advantages over permanent magnets because the magnetic fields can be turned on and off. Also, electromagnets can be varied in strength by varying the electric current.

P28.4

Magnetic fields around magnets

Collins
Secure Science

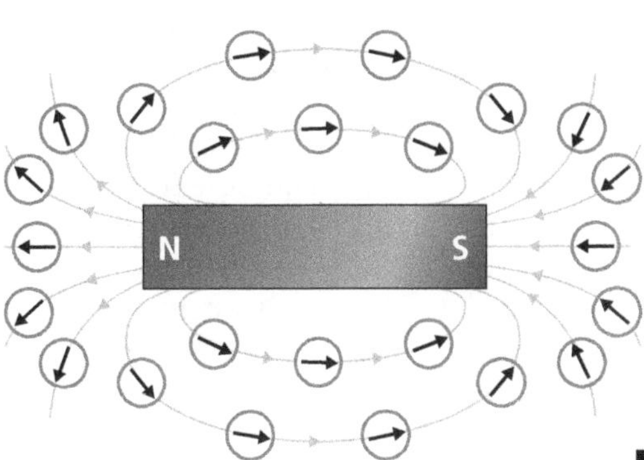

P28.5

Magnetic fields around magnets

Collins
Secure Science

A magnetic field is a region around a magnet where another magnet (or an iron/steel object) experiences a force. **Magnetic field lines** can be identified using a plotting compass. Magnetic field lines can be drawn using arrowed lines that point away from north poles and towards south poles. Magnetic fields are strongest where the magnetic field lines are closest together, i.e. at the poles. Magnetic fields get weaker as the distance away from the magnet increases. The Earth has a permanent magnetic field. A magnetic compass has a magnetised needle, and one end of the needle always points towards the Earth's north pole.

P28.5

Session 29: Atoms and radioactivity

Collins Secure Science

Atoms and radioactivity

P29.1

Session 29: Atoms and radioactivity

Atoms are the tiny, basic building blocks of **matter**. The nuclear model of the atom is a small nucleus containing protons and neutrons, surrounded by orbiting electrons. **Isotopes** of elements have the same number of protons but different numbers of neutrons. Some atoms are unstable and **emit** nuclear radiation in a random process called **radioactive decay**. Nuclear **radiation** is hazardous to living cells and precautions need to be taken when handling **radioactive** substances.

The atomic model

Collins Secure Science

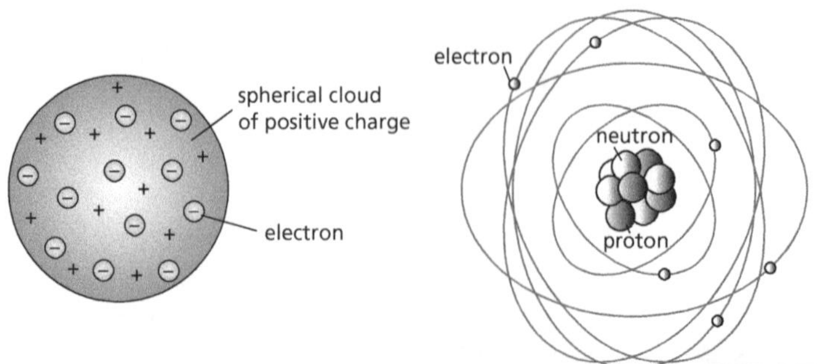

P29.2

The atomic model

Our model of the atom has changed over time. The diagram on the left shows the 'plum pudding' model from 1904. Alpha particle scattering experiments in 1911 showed that this model was incorrect, and that atoms have a tiny nucleus surrounded by electrons. This model was adapted further in 1913 to include the electrons orbiting the nucleus, as shown in the right-hand diagram. Protons were discovered in 1917 and neutrons in 1932. New experimental evidence often leads to a scientific model being changed or replaced.

Nuclear equations

Collins Secure Science

$$^{238}_{92}\text{U} \xrightarrow{\alpha\text{-decay}} {}^{4}_{2}\text{He} + {}^{234}_{90}\text{Th} \qquad (238 - 4;\ 92 - 2)$$

$$^{15}_{6}\text{C} \xrightarrow{\beta\text{-decay}} {}^{0}_{-1}\text{e} + {}^{15}_{7}\text{N} \qquad (15 - 0;\ 6 + 1)$$

P29.3

Nuclear equations

The diagram on the left represents an example of alpha particle radioactive decay. In this case, a nucleus of uranium-238 decays via the emission of an alpha particle, $^{4}_{2}\text{He}$, giving a daughter nucleus of thorium-234. During beta particle radioactive decay (diagram on the right), a nucleus of carbon-15 emits a beta particle, $^{0}_{-1}\text{e}$, leaving a daughter nucleus of nitrogen-15. In general:

$$\alpha \text{ decay } {}^{A}_{Z}\text{X} \rightarrow {}^{4}_{2}\text{He} + {}^{A-4}_{Z-2}\text{Y}$$

$$\beta \text{ decay } {}^{A}_{Z}\text{X} \rightarrow {}^{0}_{-1}\text{e} + {}^{A}_{Z+1}\text{Y}$$

Radioactive decay

Collins Secure Science

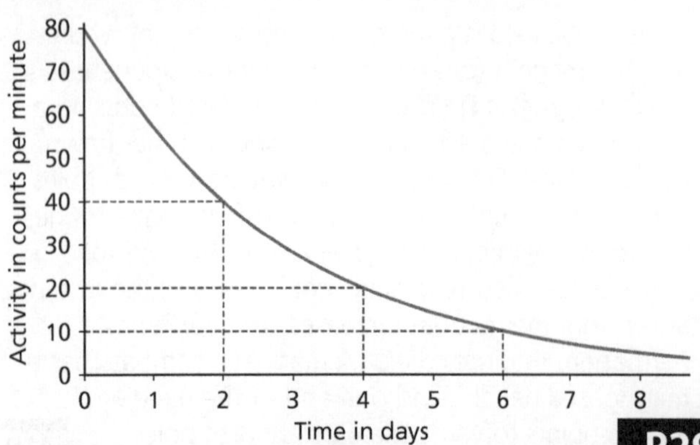

P29.4

Radioactive decay

Although radioactive decay is random, the pattern is very predictable. The **half-life** of a radioactive substance is the time taken for the number of radioactive nuclei (or the activity) to fall to half its initial value. The half-life of an isotope is always constant, but different isotopes can have different half-lives varying from tiny fractions of a second to many trillions of years. In the diagram, the half-life is 2 days. Radioactive contamination of materials occurs when radioactive substances combine with other materials. Irradiation is the process of exposing an object to nuclear radiation.

Nuclear radiation

Collins
Secure Science

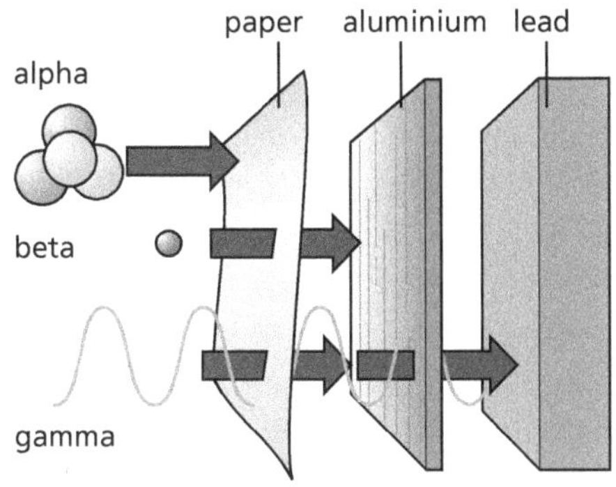

P29.5

Nuclear radiation

Collins
Secure Science

During radioactive decay, unstable nuclei can emit radiation and become stable. The rate of radioactive decay is called activity (measured in **becquerel**, Bq), and can be measured using a Geiger–Muller tube. The diagram shows three of the four types of nuclear radiation. **Alpha particles** are helium nuclei and are absorbed by a thin piece of paper. **Beta particles** are high energy electrons emitted from the nucleus. Beta particles are absorbed by a few millimetres (mm) of aluminium. Gamma rays are electromagnetic radiation emitted by the nucleus and are absorbed by a few centimetres (cm) of lead. Neutrons of elements can also be emitted by nuclei.

P29.5

Isotopes

Collins
Secure Science

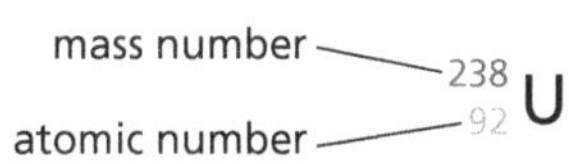

P29.6

Isotopes

Collins
Secure Science

Isotopes of elements are atoms with the same number of protons but different numbers of neutrons. The nuclei of different isotopes can be represented by the symbols $^{A}_{Z}X$, where A is the mass number, which is equal to the proton number + neutron number; Z is the atomic number, or the proton number; and X is the chemical symbol. The diagram shows the isotope uranium-238 – it has 92 protons and 146 neutrons. An isotope of lithium can be represented as $^{7}_{3}Li$ – it has 3 protons (the atomic number); and 4 neutrons in its nucleus, so the mass number is 7.

P29.6

Investigating specific heat capacity

Collins
Secure Science

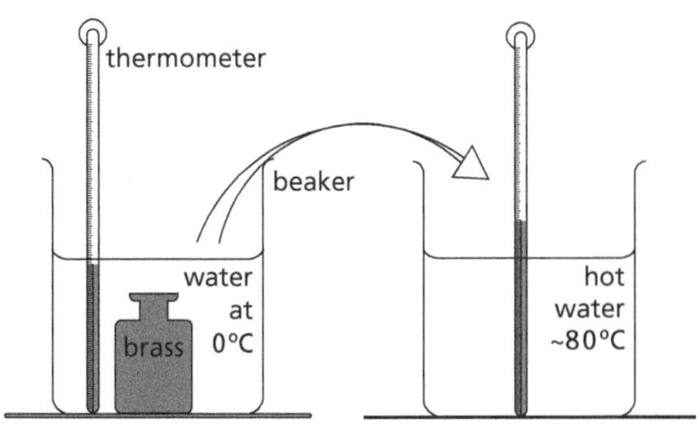

P30.1

Investigating specific heat capacity

Collins
Secure Science

In the example shown, some of the thermal energy stored in the water transfers to thermal energy stored in the brass block. This happens until the temperatures of the block and the water are the same (equilibrium), so:

thermal energy increase of brass = thermal energy decrease of water

$$\text{mass}_{\text{brass}} \times \text{specific heat capacity}_{\text{brass}} \times \text{temperature increase}_{\text{brass}} = \text{mass}_{\text{water}} \times \text{specific heat capacity}_{\text{water}} \times \text{temperature decrease}_{\text{water}}$$

Accurate and safe measurements of the water temperature and time need to be recorded. The specific heat capacity of the brass can then be calculated from the measurements. We know the specific heat capacity of water. The thermal energy wasted to the surroundings is assumed to be zero. If the wasted energy is not zero, the value calculated will not be valid.

P30.1

Using circuit diagrams

Collins
Secure Science

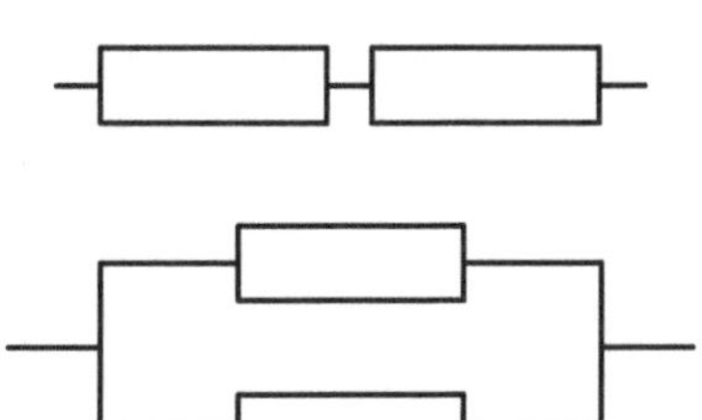

P30.2

Using circuit diagrams

Collins
Secure Science

A circuit is set up to measure the potential difference and current for different lengths of a metal wire. The equation:

resistance = potential difference / current

is used to calculate the resistance at each length, and a graph of resistance (*y*-axis) against length (*x*-axis) is plotted. The relationship between length and resistance is determined from the graph. The wire is replaced by two identical resistors in series, and then in parallel. Potential difference and current readings are then taken, and the overall resistance of each combination calculated. These are then compared with the individual resistance of the resistors.

P30.2

Investigating density

Collins
Secure Science

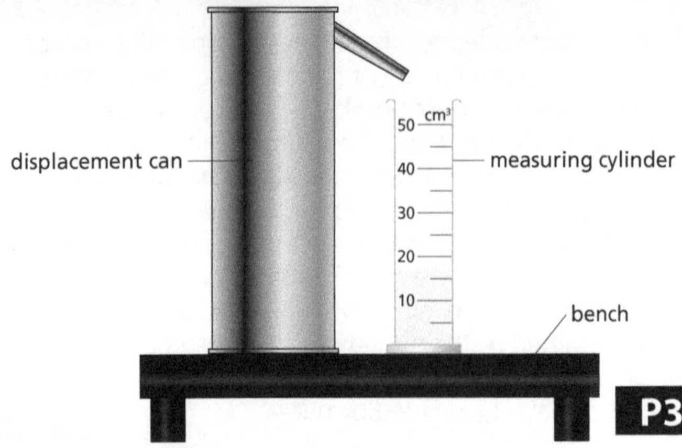

P30.3

Investigating density

Coll
Secure S

To measure density you need to know the mass and volume of an object. Objects that have a regular shape can be measured to find their volume. Irregularly shaped objects are difficult to measure, so you can find their volume using a displacement can. The volume of the regular object is calculated using:

volume = length × width × height

The measurements can then be used to calculate the density of the objects using:

density = mass / volume

P3(

Investigating acceleration

Collins
Secure Science

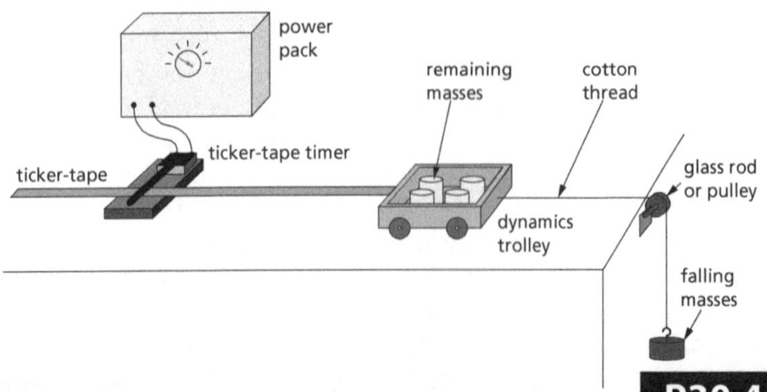

P30.4

Investigating acceleration

Coll
Secure S

A ticker-tape timer can be used to measure the acceleration of a dynamics trolley. The trolley is pulled by different falling masses that each produce a different pulling force. This investigates the link between force and acceleration. Masses are moved from the trolley to the falling mass stack. This varies the pulling force but keeps the overall mass the same. Adding more masses to the trolley but keeping the pulling force constant, investigates the link between mass and acceleration. The results of the investigation can be compared with Newton's Second Law.

P3(

Investigating IV characteristics

Collins
Secure Science

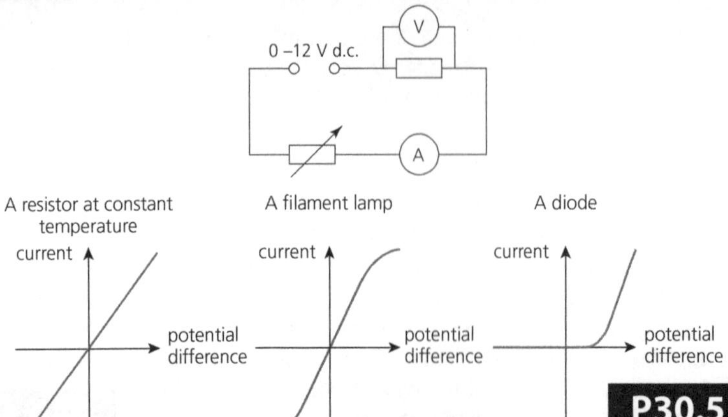

P30.5

Investigating IV characteristics

Coll
Secure S

The three variables are: the component being tested; the potential difference, *V* (that you change); and the current, *I* (th you measure). The diagram shows the circuit needed to make the measurements. The potential difference is changed in ste over a range using the variable resistor. Reversing the connect on the power supply gives negative values. Then use the average values to plot the **IV** characteristics. A graph of *I* (*y*-ax against *V* (*x*-axis) can then be plotted for each component. The **IV** graphs, called characteristics, should look like the ones shown in the diagrams. Uncertainties can be reduced by switching off the power between readings. Repeat the readings and find the mean (average). Then use the mean values to plot the **IV** characteristics.

P3(

Investigating the behaviour of springs

Collins
Secure Science

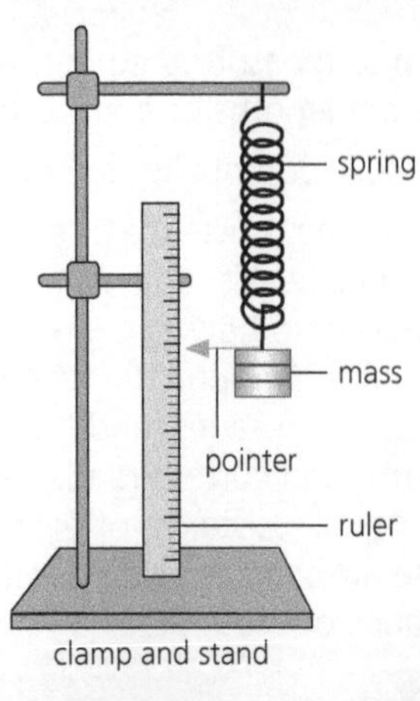

P30.6

Investigating the behaviour of springs

Coll
Secure S

The independent variable is the force on the spring (produced by changing the mass). The dependent variable is the **extension** of the spring. The main control variable is keeping the spring the same. It is important to measure the extension of the spring NOT the overall length of the spring. Too much force should not be put on the spring, otherwise it will permanently stretch and produce an **uncertainty** in the measurements. A force–extension graph should then be plotted, and the pattern noted.

We know the specific heat capacity of water. The thermal energy wasted to the surroundings is assumed to be zero. If the wasted energy is not zero, the value calculated will not be valid.

P3(

Investigating waves in a ripple tank and on a string

Collins Secure Science

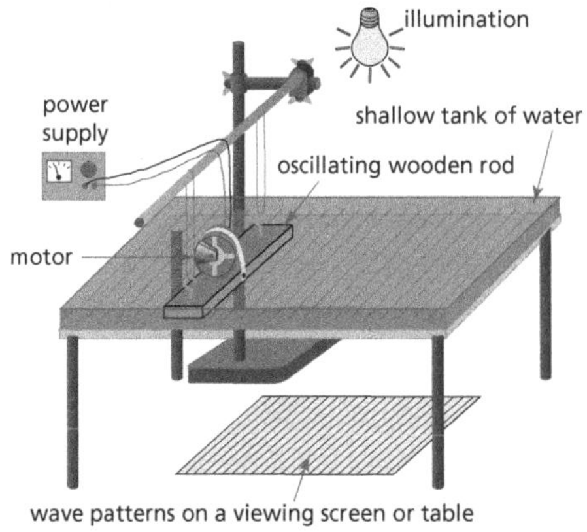

P30.7

Investigating waves in a ripple tank and on a string

Collins Secure Science

For the ripple tank, a strobe can 'freeze' the pattern of the waves seen on the viewing screen. The frequency of the waves can be read off the strobe and the wavelength is measured accurately with a ruler. The speed of the waves can be calculated using:

wave speed = frequency × wavelength

P30.7

Investigating waves in a ripple tank and on a string

Collins Secure Science

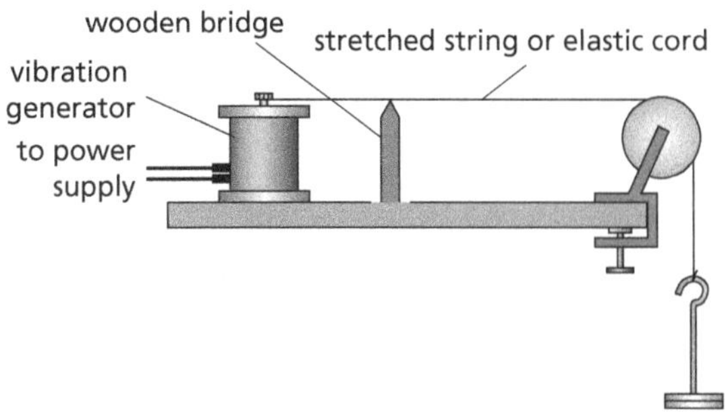

P30.8

Investigating waves in a ripple tank and on a string

Collins Secure Science

The strobe can also 'freeze' the motion of the stationary waves produced on the vibrating string. The frequency of the waves is measured from the vibration generator power source. The wavelength is measured directly using a ruler.

P30.8

Physical skills

Collins Secure Science

Measuring forces – forces are normally measured using a newtonmeter, which is also known as a spring-balance. A force extends a spring, and the position of a pointer attached to the spring is read off a scale to determine the force. Different force ranges can be measured by using newtonmeters with springs with different stiffnesses. Newtonmeters can have a hook on the end to measure tension (pull) forces, or they can have a flat pad to measure compressions (pushes).

P30.9

Physical skills

Collins Secure Science

Measuring speed – speed requires the measurement of a distance (using a ruler or a tape measure) and the measurement of a time (using a stopwatch). The speed is calculated using the equation:

$$\text{speed} = \frac{\text{distance}}{\text{time}}$$

Instantaneous speed is the speed of an object at any given point during its motion; mean speed is the mean speed over the course of the motion. A 100 m sprinter can run at a maximum instantaneous speed of about 12 m/s, but their mean speed over the whole 100 m is about 9 m/s.

P30.9

Physical skills

Collins Secure Science

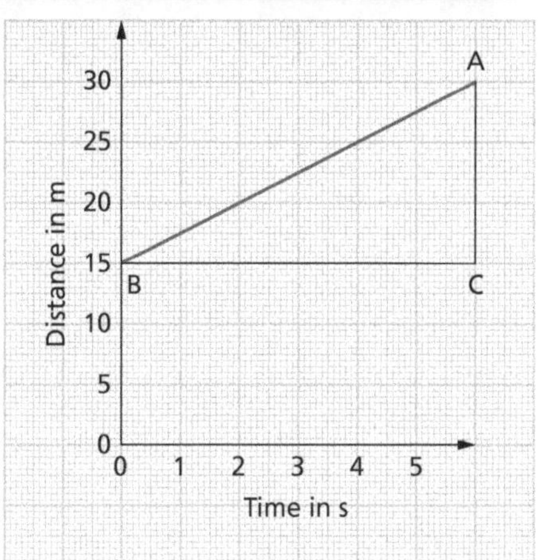

P30.10

Physical skills

Collins Secure Science

Calculating from graphs – there are several quantities that can be measured from a graph. The gradient (or slope) of a straight line can be determined by measuring the change in *y*-axis values of the line and the change in *x*-axis values of the line. The gradient is determined by:

$$\text{gradient} = \frac{\text{change in } y\text{-axis values}}{\text{change in } x\text{-axis values}}$$

In the graph shown

$$\text{gradient} = \frac{AC}{BC}$$

Values can also be read from lines of best fit. It is always best to use a ruler to draw horizontal and vertical lines from the line of best fit. You can then read off the values from the scales on the axes.

P30.10

Physical skills

Collins
Secure Science

Hazards in physics – some of the required practicals in physics involve hazards. The specific heat capacity experiment involves using hot water, which requires the use of safety goggles. Hot objects should be allowed to cool before handling. During the electrical experiments, if mains power supplies are being used, these should be kept away from water taps and sinks. The density experiment has no obvious risks. The acceleration and the spring experiments both involve the use of weights, which could fall onto feet. Feet should be kept out of the way of potential falls. The ripple tank experiment also potentially involves using a mains power supply and water. These should be kept far apart. The infrared experiments involve a mains infrared heater and a Leslie cube, both of which can get very hot. These should be allowed to cool before handling.

P30.11

Physical skills

Coll
Secure S

Using equations to explain observations – in several of the required practicals, you are asked to investigate the relationship between a dependent and an independent variable that are linked through an equation. Examples of this are:

potential difference = current × resistance

resultant force = mass × acceleration

force applied to a spring = spring constant × extension

wave speed = frequency × wavelength

Each of these can produce a graph with a straight line. If the data values collected form a straight line on the graph, then the equation has been proved.

P30.

Investigating infrared radiation

Collins
Secure Science

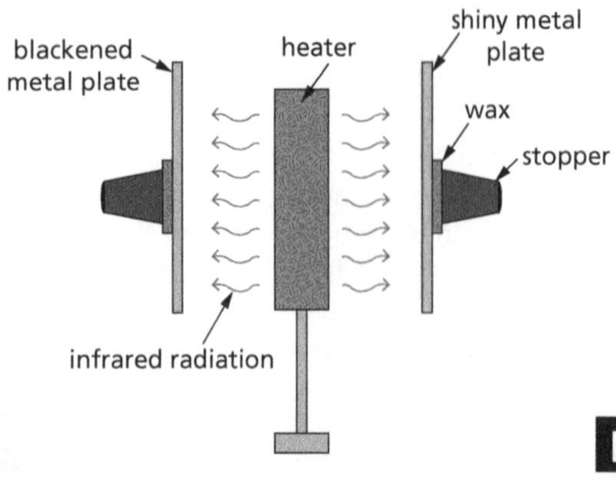

P30.12

Investigating infrared radiation

Coll
Secure S

The absorption of infrared radiation by different surfaces is compared using the apparatus shown. The time taken for each stopper to fall off the plates is recorded. The best absorber heats up fastest and the wax melts quickest. The controlled variables are the distance from the heater to the plates and the plate area.

P30

Investigating infrared radiation

Collins
Secure Science

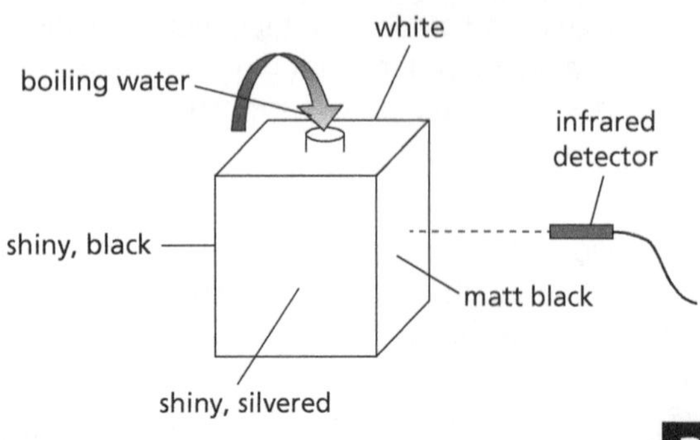

P30.13

Investigating infrared radiation

Coll
Secure S

An infrared detector measures the intensity of infrared radiation emitted from each side of the Leslie cube. The experiment must be performed quickly to ensure the water is the same temperature for each reading. The distance between each face and the detector must also be controlled.

P30

Measuring liquids

Collins Secure Science

Measuring cylinders

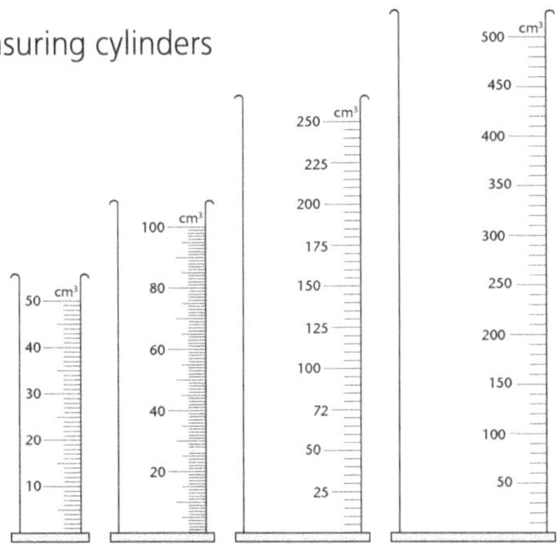

A1

Measuring liquids

Collins Secure Science

Used for measuring the volume of liquids.

Tip: Choose the smallest measuring cylinder for the volume you want to measure to minimise uncertainty.

A1

Measuring liquids

Collins Secure Science

Displacement (Eureka) can

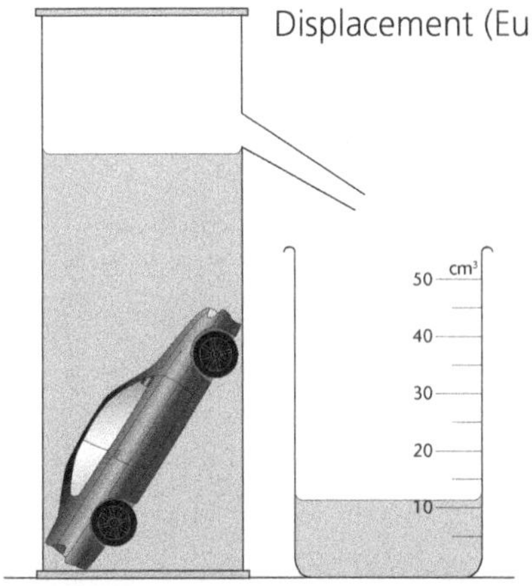

A2

Measuring liquids

Collins Secure Science

A container with a spout at the top, that is used to find the volume of an irregularly shaped object.

Tip: The accuracy of your measurement relies on the displacement can being full before the object is added. Fill the displacement can where you are going to use it to avoid losing water when you move it.

A2

Measuring liquids

Collins Secure Science

Syringe

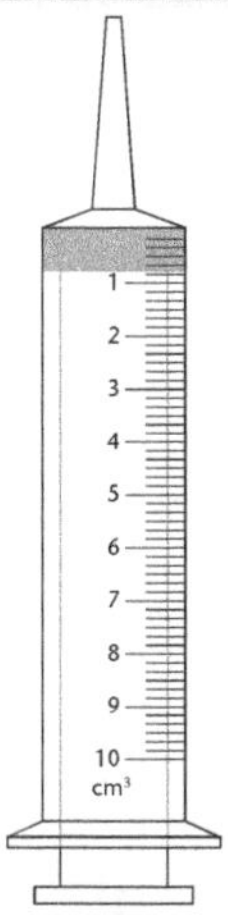

A3

Measuring liquids

Collins Secure Science

Used for accurately measuring small volumes of liquids.

Tip: Make sure you have removed any bubbles from the liquid to get an accurate volume of liquid.

A3

Measuring gases

Collins Secure Science

Delivery tube with bung

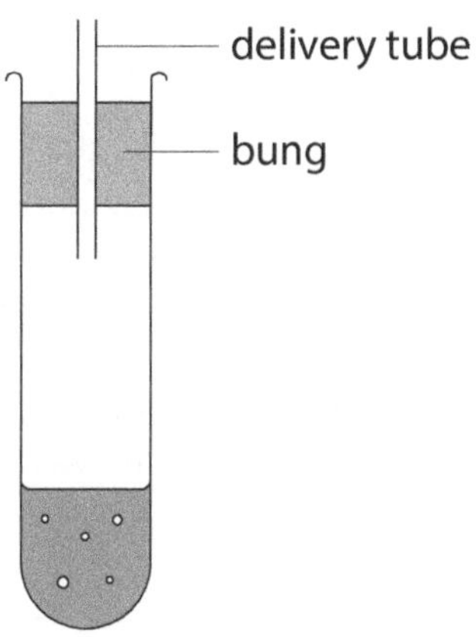

A4

Measuring gases

Collins Secure Science

Transfers a gas from where it is made to a container where it is collected.

Tip: Make sure that the bung fits snugly otherwise gas will escape.

A4

Measuring gases

Collins Secure Science

Inverted test tube, filled with water

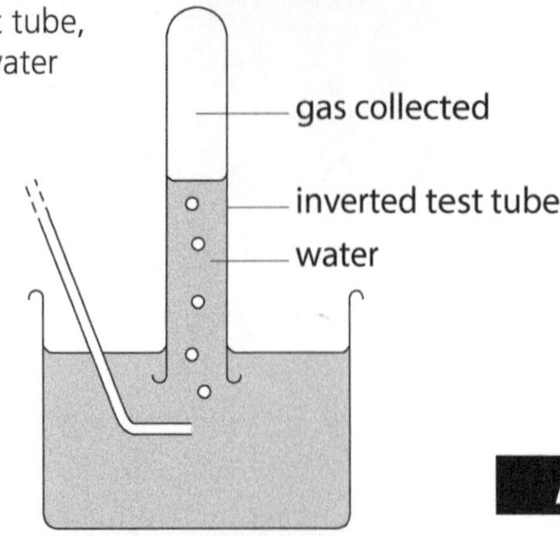

A5

Measuring gases

Can be used to collect gas that does not dissolve in water.

Tip: It is tricky to get the test tube in the right place without bubbles of air. Use a large trough so that you can invert the test tube and put the delivery tube into it underwater.

Measuring gases

Collins Secure Science

Gas syringe

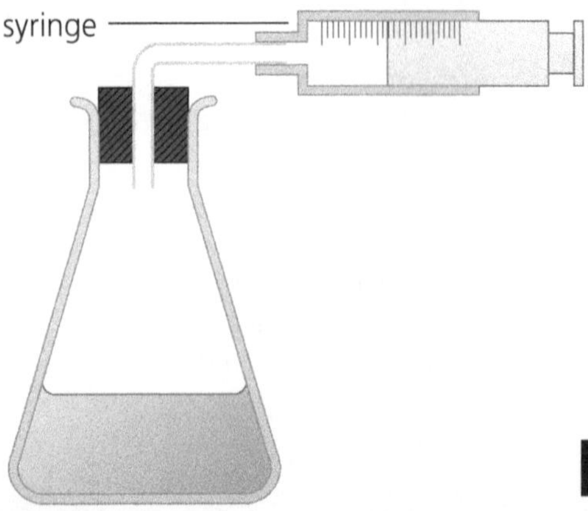

A6

Measuring gases

Can be used to collect gases that will dissolve in water.

Tip: Check the syringe has a good seal before you start to make sure gas is not lost.

Containers

Collins Secure Science

Conical flasks

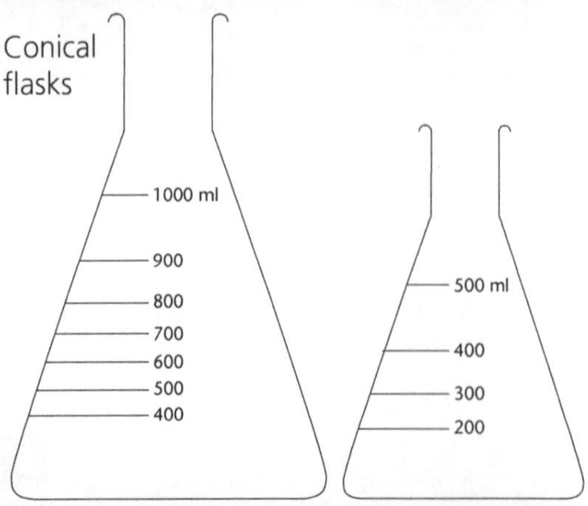

A7

Containers

Container that is specially shaped for swirling reactants.

Tip: Keep the conical flask as level as possible when you swirl to mix the reactants to avoid spillages.

Containers

Collins Secure Science

Beakers

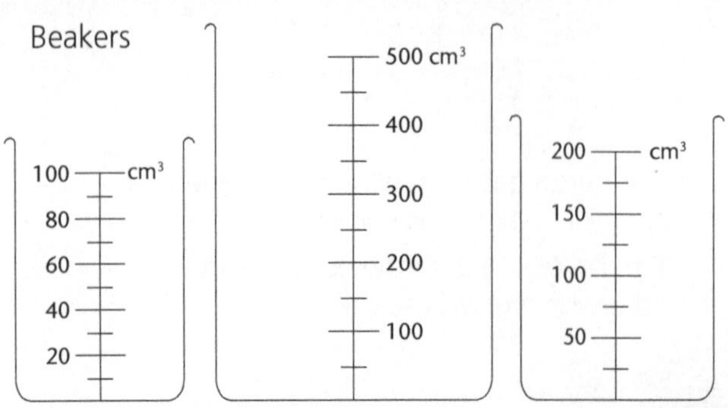

A8

Containers

An open container for pouring liquids, reacting substances and heating liquids.

Tip: Beakers are easily damaged – take care when mixing with glass or metal stirrers, and heat carefully to avoid breakages. Heating and cooling quickly will cause beakers to crack so do these things slowly.

Containers

Collins Secure Science

Boiling tube

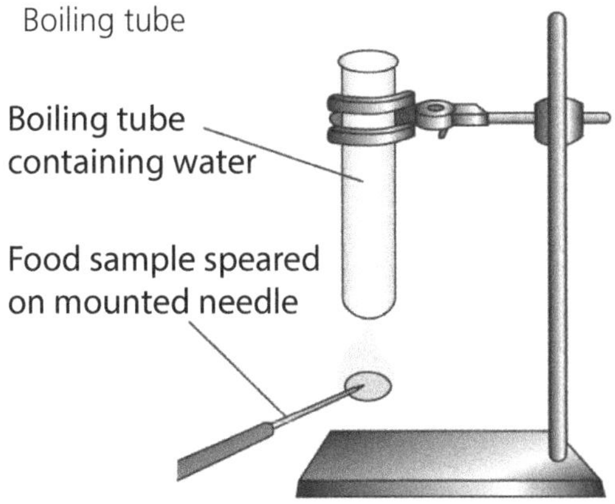

A9

Containers

Collins Secure Science

An open tube for strongly heating substances and reacting around 20 cm^3 of reactants.

Tip: Substances can easily bubble out of a boiling tube when reacting or heating so make sure the boiling tube is pointing away from people. Always handle with a holder, especially when hot.

A9

Containers

Collins Secure Science

Test tube

A10

Containers

Collins Secure Science

A small open tube for reacting around 10 cm^3 of reactants.

Tip: Substances can easily bubble out of a test tube when reacting or heating so make sure the test tube is pointing away from people. Always handle with a holder, especially when hot.

A10

Containers

Collins Secure Science

Test tube and boiling tube rack

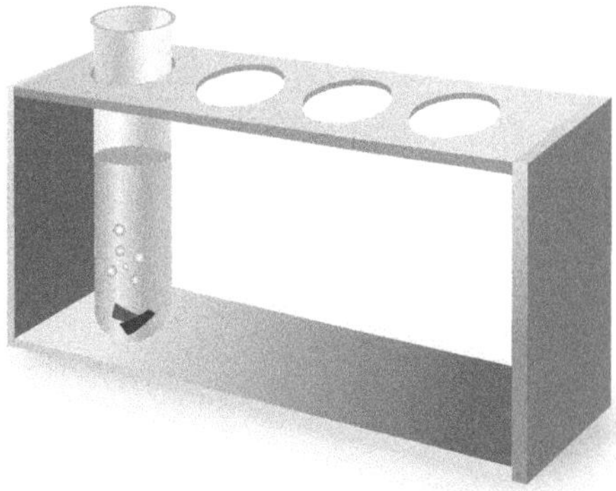

A11

Containers

Collins Secure Science

A stand for holding test tubes or boiling tubes.

Tip: Test tubes and boiling tubes can easily roll off the bench and break, so always use a rack to hold them, even when not being used.

A11

Containers

Collins Secure Science

Dropping bottle

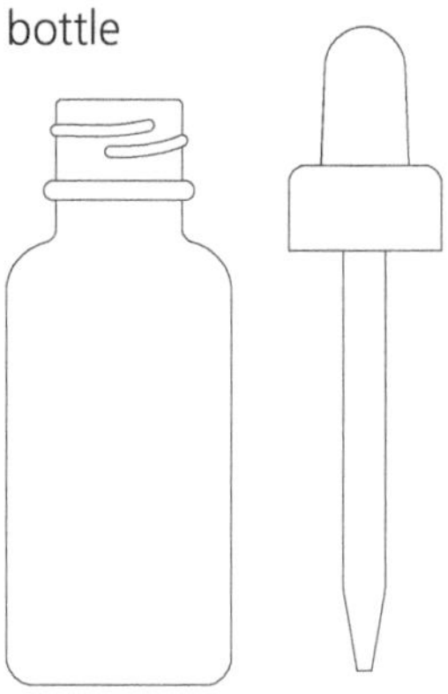

A12

Containers

Collins Secure Science

A glass bottle with a dropper in the lid, that is used to place drops of the substance into another container.

Tip: It is easy to contaminate the contents of dropper bottles so make sure the dropper does not touch anything else. The lids of dropper bottles are not attached to the rest of the bottle so always pick them up by the bottle and never by the lid.

A12

Containers

Expanded polystyrene cup and lids

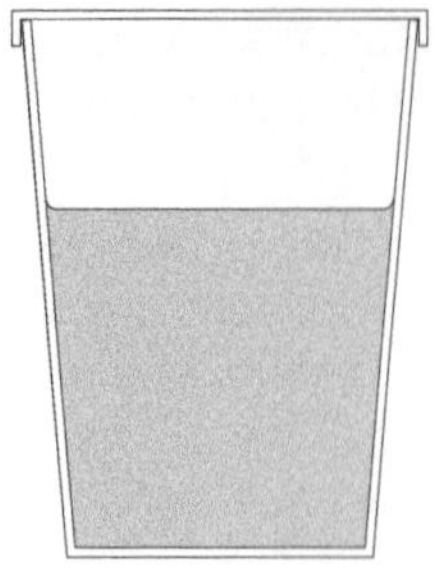

A13

Containers

An open plastic cup with a lid that is used to investigate the cooling of liquids or temperature changes during reactions.

Tip: Polystyrene is not biodegradable and paper cups function just as well.

A1

Containers

Leslie cube

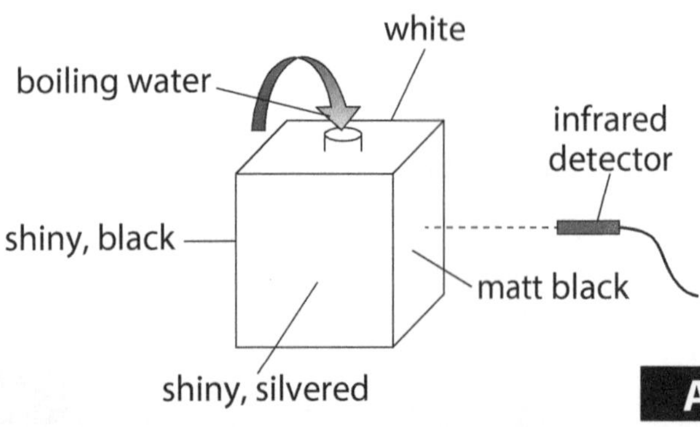

A14

Containers

A metal container that is filled with hot water and has surfaces that are different colours and textures. Used to investigate how heat energy is transferred by thermal radiation.

Tip: The Leslie cube needs to be filled with a large volume of very hot water so take care to avoid burns. Make sure the cube is cooled completely before moving or emptying.

A1

Containers

Spotting tile

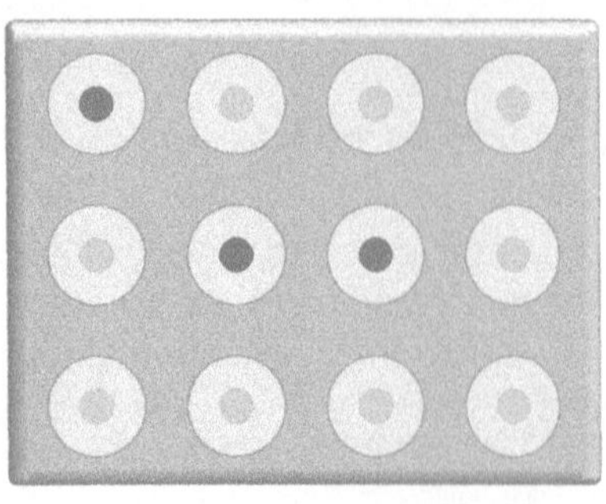

A15

Containers

A plastic tile with wells that hold a small volume of liquid for analysis.

Tip: It is easy to contaminate the wells, so make sure the tile is clean before you start. Be very careful to only drop substances into one well at a time.

A1

Containers

Capillary tube

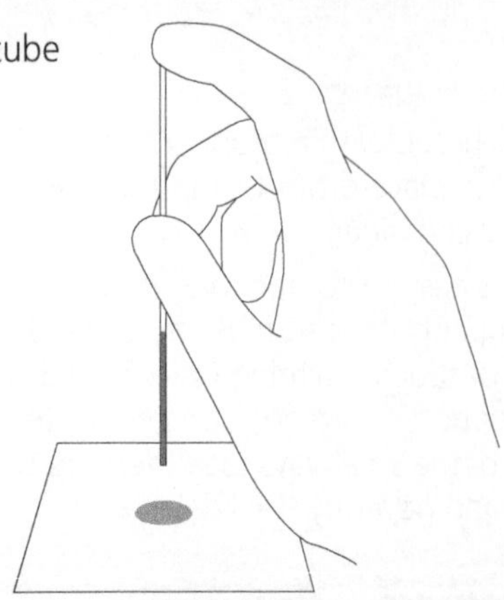

A16

Containers

Tiny glass tube used to move tiny amounts of a liquid.

Tip: Capillary tubes are very delicate so handle with care to avoid snapping them. Never leave capillary tubes on the bench where they could roll off. It is very difficult to clean capillary tubes so use a new one every time.

A1

Measuring length and area

Collins Secure Science

Ruler

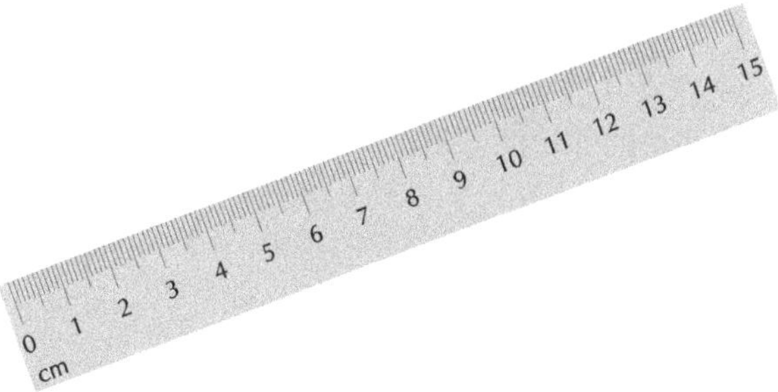

A17

Measuring length and area

Collins Secure Science

Used for drawing straight lines, measuring lengths up to 30 cm, and can be used to measure reaction times when ruler is dropped.

Tip: When drawing lines of best fit and axes on a graph, use a ruler that covers the whole distance, and a sharp pencil.

A17

Measuring length and area

Collins Secure Science

Metre rule

A18

Measuring length and area

Collins Secure Science

Used for measuring lengths up to a metre.

Tip: Plastic metre rulers are not as robust as wooden metre rulers so should be treated with more care.

A18

Measuring length and area

Collins Secure Science

Tape measure

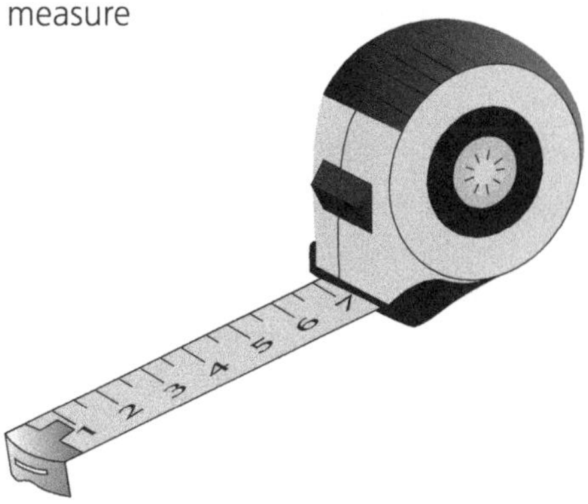

A19

Measuring length and area

Collins Secure Science

Used for measuring large distances up to 30 m, such as an area of a field.

Tip: It can be tricky to measure accurately if working alone – make sure you secure the end of the tape measure or have someone hold it in position.

A19

Measuring length and area

Collins Secure Science

Quadrat

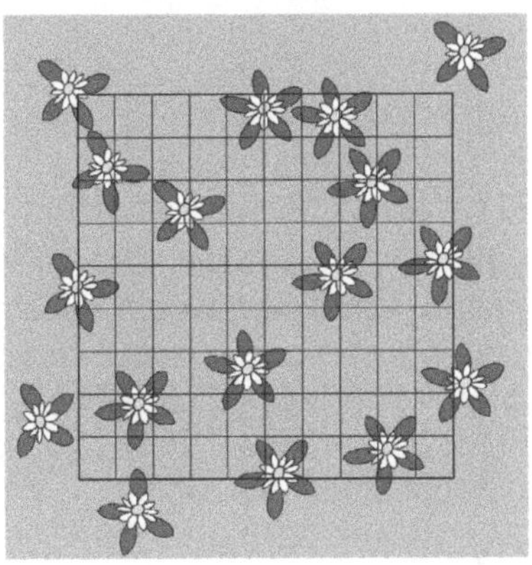

A20

Measuring length and area

Collins Secure Science

An outline of a square used to estimate the coverage of organisms that do not move. Some quadrats have an internal grid.

Tip: Quadrats must be placed randomly and measurements repeated several times to get a fair estimation of the coverage of the organisms.

A20

Separation

Collins
Secure Science

Chromatography paper

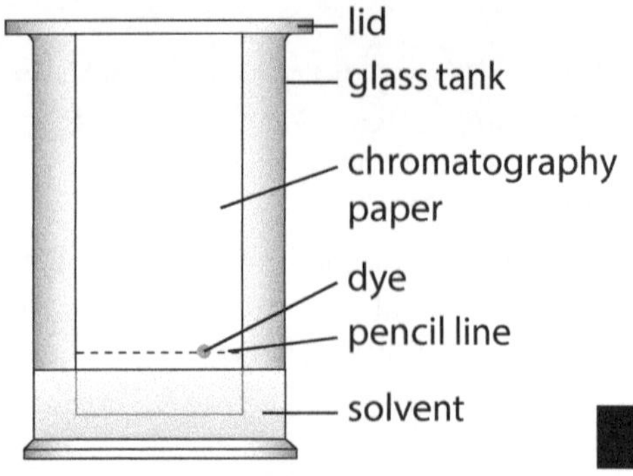

A21

Separation

Coll
Secure S

Thick paper that allows substances with different solubilities to separate out by dissolving in a solute moving through the paper.

Tip: Make sure you do not dip the substance to be separated into the solution or it will wash away. Use pencil to draw the starting point line – ink could dissolve and spread over the paper.

A2

Separation

Collins
Secure Science

Filter funnel and filter paper

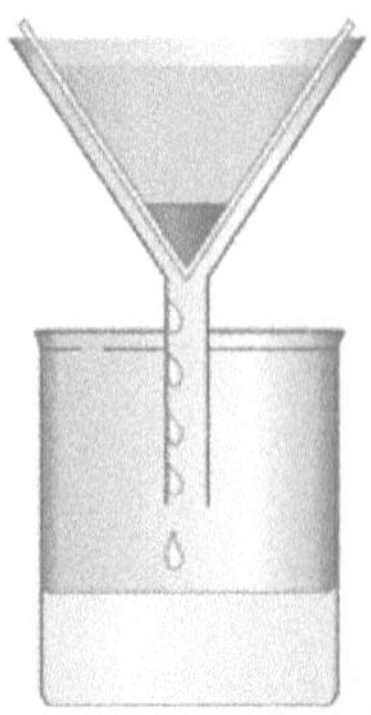

A22

Separation

Coll
Secure S

Separates insoluble substances from a liquid.

Tip: Make sure all the liquid runs through the paper and not down the side of the funnel. Fold the circle of filter paper in half, and then in half again, then open out one leaf of the paper to make a cone. Do not overfill the paper – you may need to be patient and wait for some to drain out.

A2

Separation

Collins
Secure Science

Crystallising dish

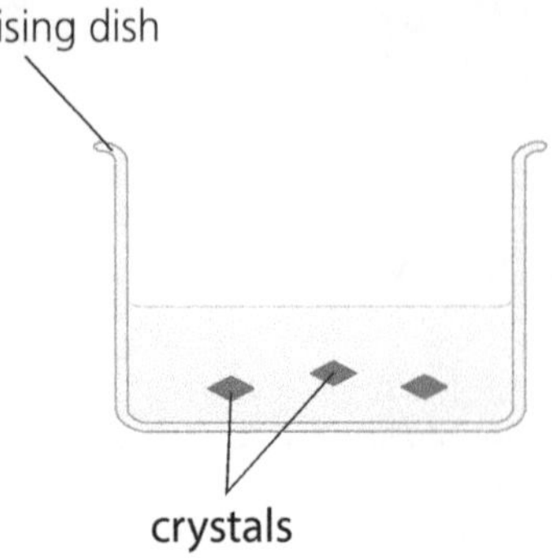

A23

Separation

Coll
Secure S

A short, open glass dish that allows liquid to evaporate to leave crystals behind in the dish.

Tip: Crystallising dishes are not designed for heating – use an evaporating basin for heating.

A2

Separation

Collins
Secure Science

Evaporating basin

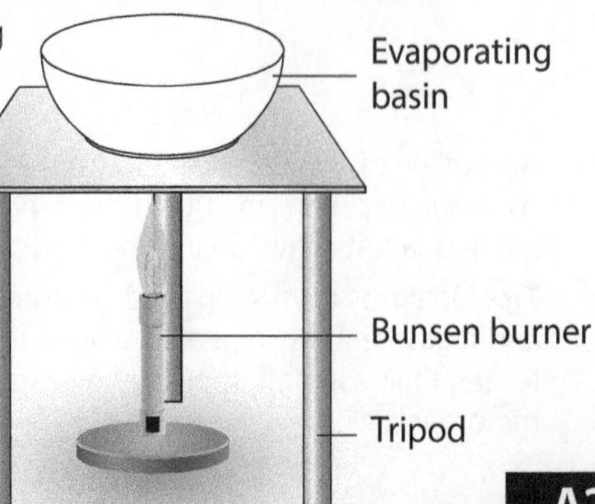

A24

Separation

Coll
Secure S

An open dish made of porcelain that can be heated to evaporate water from a solution, leaving crystals of the solute behind.

Tip: Make sure you use the right flame for evaporating – the flame should be blue and low enough to not be in contact with the dish. Remove the heat before all the liquid has evaporated so the solid does not spit. Leave to cool before handling.

A2

Separation

Collins
Secure Science

Distillation

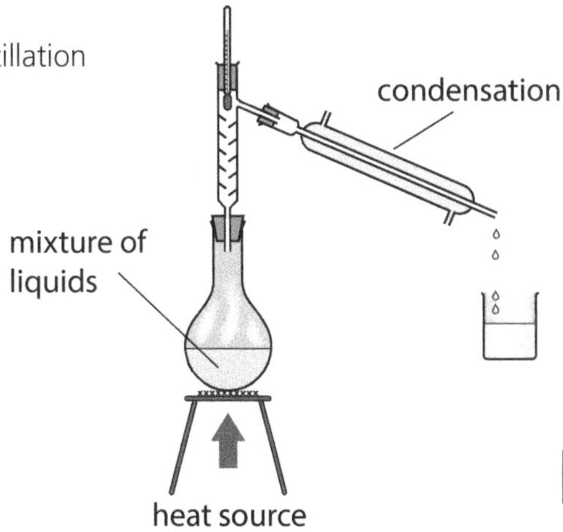

A25

Separation

Collins
Secure Science

A mixture of substances with different boiling points can be separated by heating and then condensing the first liquid to boil so it can be collected.

Tip: Anti-bumping granules are usually added to the mixture of liquids so they boil evenly. This makes the process safer and reduces the risk of losing any substances.

A25

utting and changing solid substances

Collins
Secure Science

Scalpel

A26

Cutting and changing solid substances

Collins
Secure Science

Very sharp blade mounted in a handle that is used to make small, precise cuts.

Tip: Scalpels are very sharp and their use should be considered very carefully as often scissors can do the same job with less risk.

A26

utting and changing solid substances

Collins
Secure Science

Cork borer

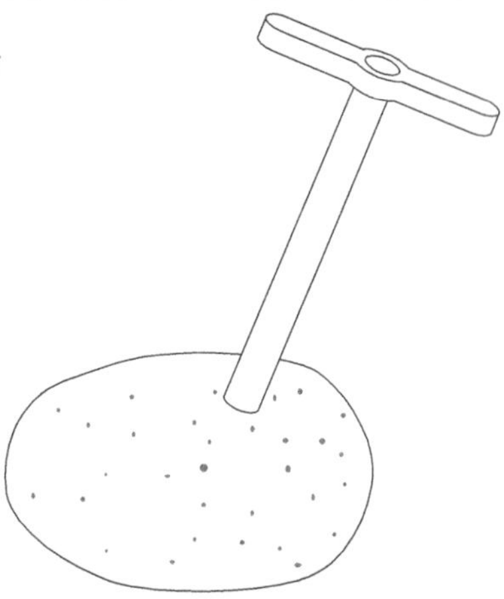

A27

Cutting and changing solid substances

Collins
Secure Science

Circular cylinder made of metal that is used to make cylinders of plant material for investigations.

Tip: Cork borers work in pairs – a large cork borer is used to cut the cylinder, and then a smaller one is used to push the bore out of the bigger cork borer.

A27

utting and changing solid substances

Collins
Secure Science

Pestle and mortar

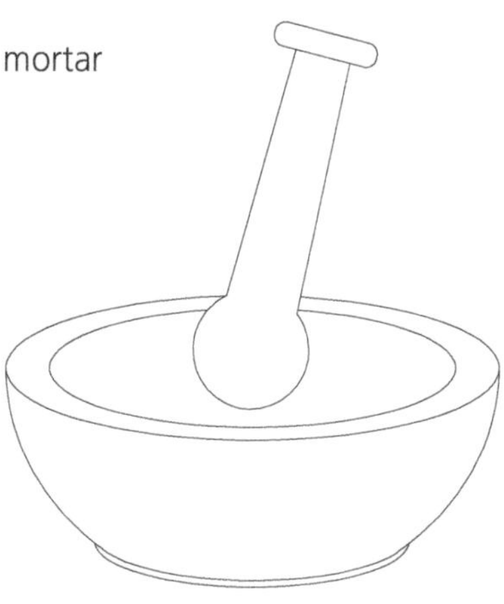

A28

Cutting and changing solid substances

Collins
Secure Science

Ceramic bowl with a wide, heavy tool used for breaking up and grinding solids.

Tip: The mortar is the bowl. This apparatus has a long history of use dating back to the stone age, and is commonly used in the preparation of pharmaceuticals.

A28

Cutting and changing solid substances

Collins
Secure Science

White tile

A29

Cutting and changing solid substances

Coll
Secure S

Provides a surface for cutting on.

Tip: White tiles prevent contamination as they are easy to clean and protect the bench from damage.

A2

Sources of energy

Collins
Secure Science

Water bath – electric

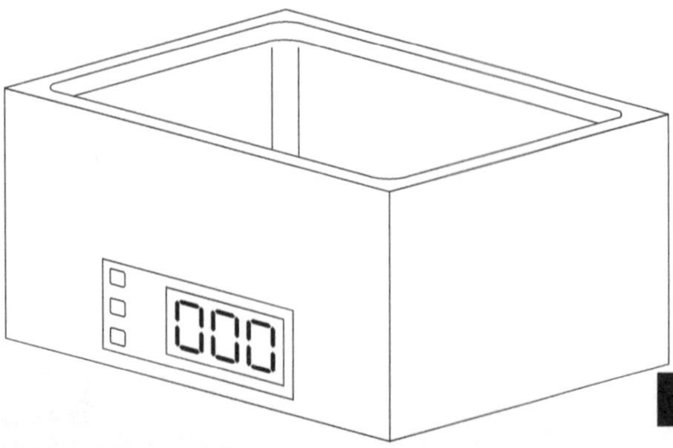

A30

Sources of energy

Coll
Secure S

An electronically heated basin of water that can be controlled with a thermostat.

Tip: Water baths can take some time to heat up so switch them on in plenty of time and use a thermometer to check the temperature throughout your investigation.

A3

Sources of energy

Collins
Secure Science

Water bath – Bunsen

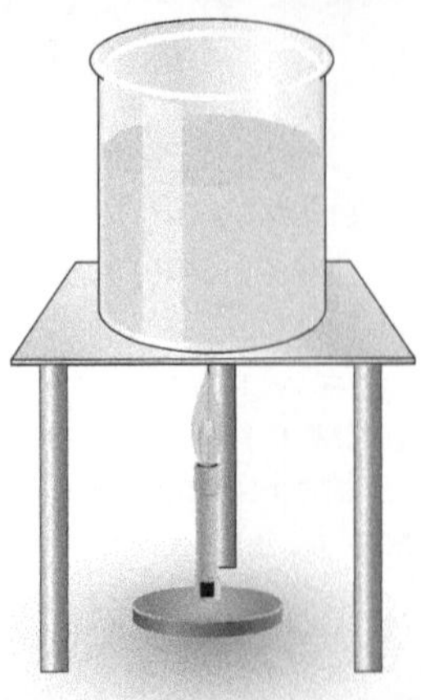

A31

Sources of energy

Coll
Secure S

A beaker of water heated with a Bunsen burner.

Tip: A water bath heated by a Bunsen will change temperature over time. Make sure you record the temperature throughout the investigation so any changes that could affect your results are noted.

A3

Sources of energy

Collins
Secure Science

Power pack

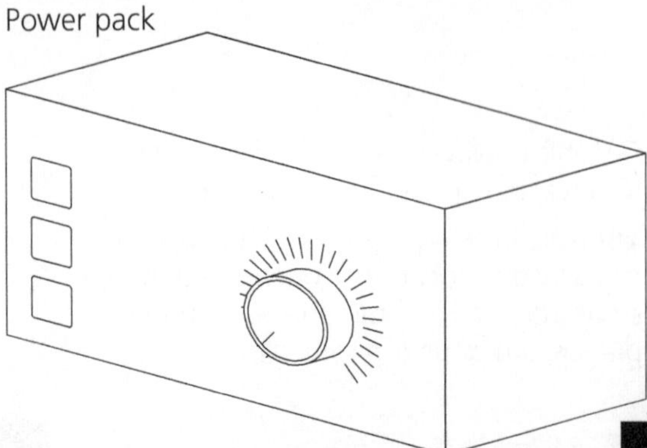

A32

Sources of energy

Coll
Secure S

Provides electrical energy for electrical circuits.

Tip: Powerpacks can provide a constant or variable potential difference depending on the investigation. Turn off when not in use.

A3

Sources of energy

Collins
Secure Science

Bunsen burner

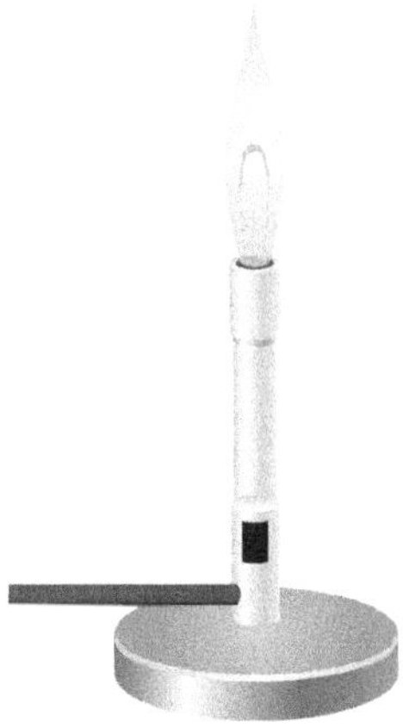

A33

Sources of energy

Collins
Secure Science

Burns methane gas to provide heat.

Tip: Only use a Bunsen to heat or burn what you have been told to heat or burn. If the flame goes out for any reason, turn it off straight away. Only turn it back on when you have a lit splint ready to light it. Check each part of a Bunsen before you use it. Never use the yellow safety flame to heat anything – it is not hot enough and will make anything that goes near it black with soot.

A33

Sources of energy

Collins
Secure Science

Ray box lamps

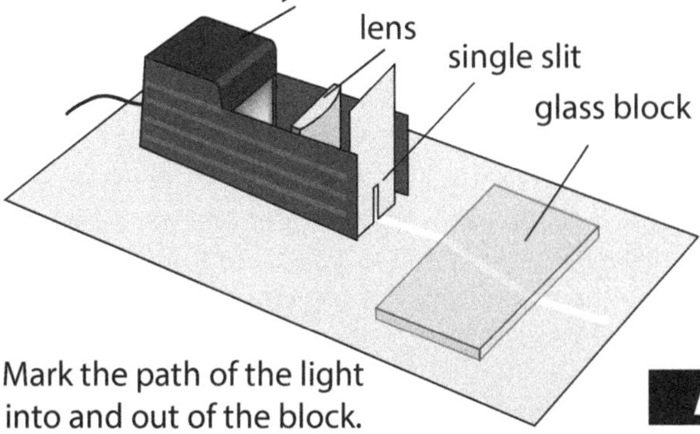

Mark the path of the light into and out of the block.

A34

Sources of energy

Collins
Secure Science

A bulb in a metal case that connects to a power pack. Allows you to create a single beam of light.

Tip: Ray boxes can get very hot – move with care.

A34

Measuring energy, resistance, waves and force

Collins
Secure Science

Ammeter

A35

Measuring energy, resistance, waves and force

Collins
Secure Science

Measures current in an electrical circuit

Tip: Connected in series with other components.

A35

Measuring energy, resistance, waves and force

Collins
Secure Science

Voltmeter

A36

Measuring energy, resistance, waves and force

Collins
Secure Science

Measures potential difference/voltage around components in an electrical circuit.

Tip: Connected in parallel across components.

A36

Measuring energy, resistance, waves and force

Collins
Secure Science

A37

Measuring energy, resistance, waves and force

Coll
Secure S

Measures the forces acting on an object

Tip: Also known as a spring balance.

A3

Measuring energy, resistance, waves and force

Collins
Secure Science

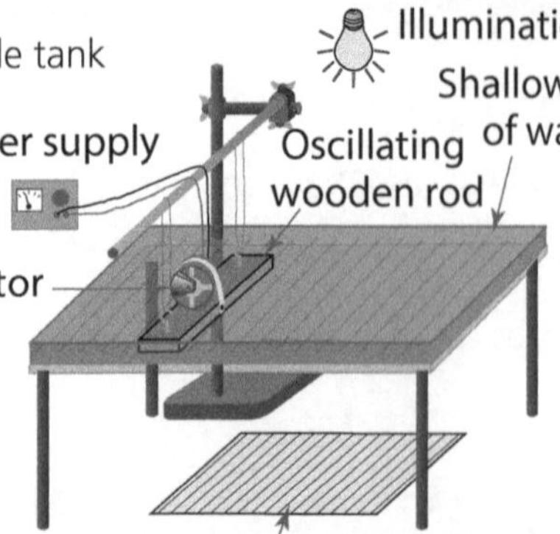

A38

Measuring energy, resistance, waves and force

Coll
Secure S

A shallow tank of water with a light above it, and an object that can make waves in it. Makes it possible to measure the properties of waves.

Tip: This apparatus brings water and electricity close together. Take care to make sure the water does not come into contact with the electricity. .

A3

Measuring energy, resistance, waves and force

Collins
Secure Science

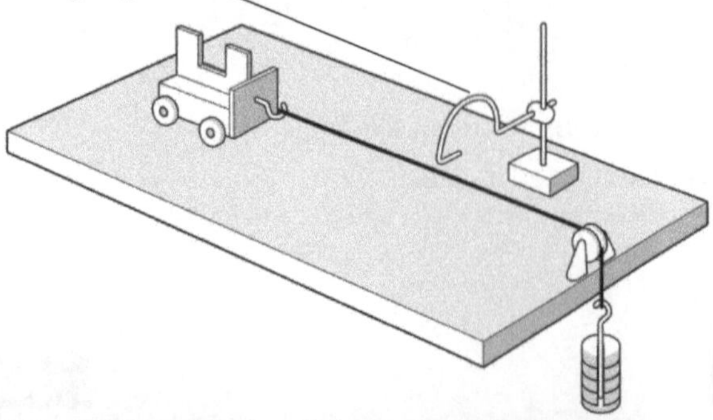

A39

Measuring energy, resistance, waves and force

Coll
Secure S

Used with computer software and connected via a datalogger. Light gates directly record the motion of objects (speeds, accelerations and time intervals).

Tip: The same datalogger and computer software system can also measure other variables, such as temperature, colour, sound, with the correct sensor.

A3

Measuring energy, resistance, waves and force

Collins
Secure Science

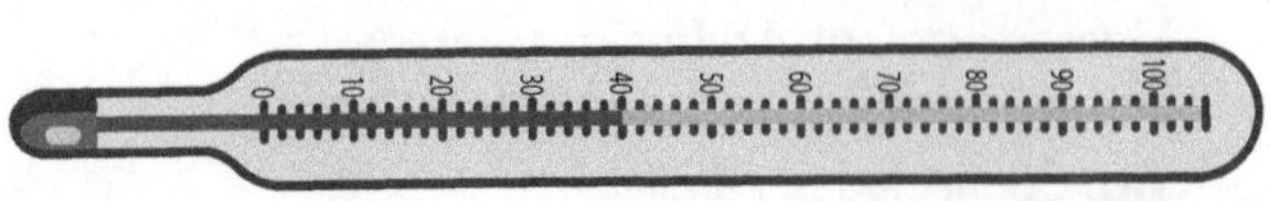

A40

Measuring energy, resistance, waves and force

Coll
Secure S

A tube usually made of glass, containing a thin column of coloured alcohol that gets bigger or smaller when the temperature changes.

Tip: Make sure you leave the thermometer in the substance you are measuring for enough time for the thermometer to show the correct temperature. Do not take the thermometer out when reading it. Thermometers are not stirring rods and will break if used in that way.

A4

Measuring energy, resistance, waves and force

Collins
Secure Science

Ticker-tape timer

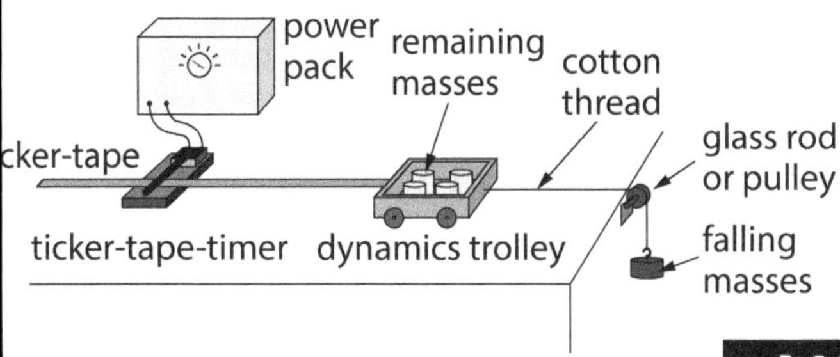

A41

Measuring energy, resistance, waves and force

Collins
Secure Science

A device for measuring and recording the motion of objects, by putting dots on a thin piece of paper.

Tip: Each dot on the paper is 0.02 s apart from the previous dot. The distance between each dot can be measured using a ruler and the speed of the object calculated.

A41

Measuring energy, resistance, waves and force

Collins
Secure Science

Dynamics trolley

A42

Measuring energy, resistance, waves and force

Collins
Secure Science

A wheeled trolley used to investigate the motion of moving objects.

Tip: Dynamics trolleys can be pulled across a bench using a falling mass-stack attached to string. The motion can be measured using a ticker-tape timer or light gates.

A42

Measuring energy, resistance, waves and force

Collins
Secure Science

Tension spring

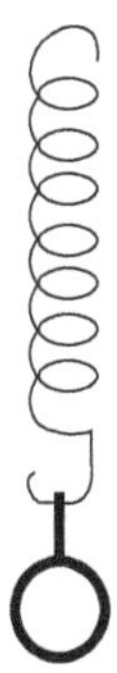

A43

Measuring energy, resistance, waves and force

Collins
Secure Science

A double-looped extension spring used to investigate the behaviour of springs.

Tip: The extension of a spring can be measured directly by arranging the zero of a ruler to be in line with the end of an unextended spring.

A43

Measuring energy, resistance, waves and force

Collins
Secure Science

Vibration generator

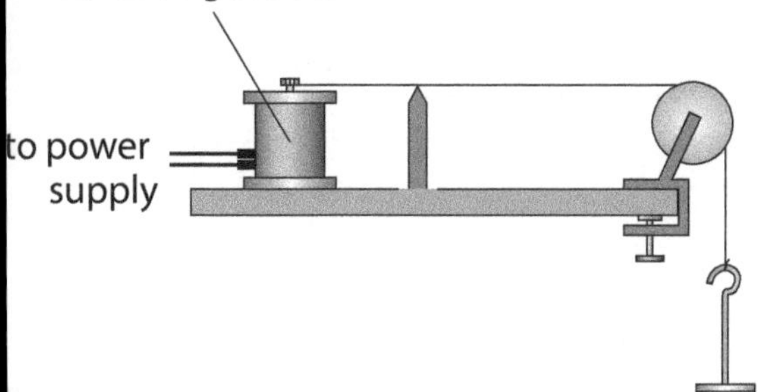

A44

Measuring energy, resistance, waves and force

Collins
Secure Science

A vibrating post that moves up and down very quickly generating waves on a supported string.

Tip: The vibration generator is powered using a frequency generator which is a power supply that produces a variable frequency potential difference.

A44

Measuring energy, resistance, waves and force

Collins
Secure Science

Bench pulley

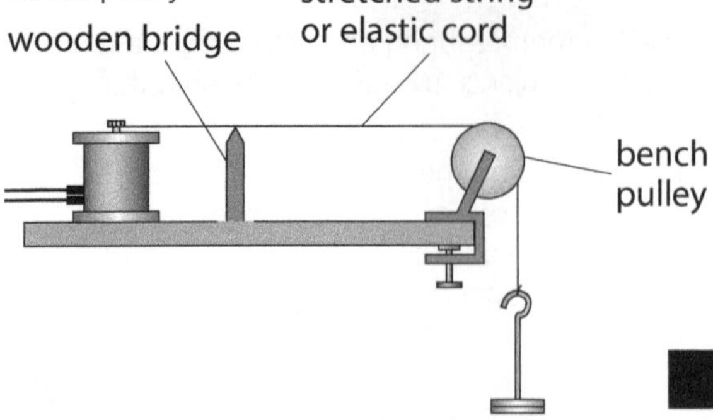

A45

Measuring energy, resistance, waves and force

Coll
Secure S

A pulley mounted on a block that can be screwed to a bench.

Tip: Bench pulleys can be used in experiments looking at the waves on a string or in experiments to pull a dynamics trolley across a bench.

A4

Measuring energy, resistance, waves and force

Collins
Secure Science

Infrared detector

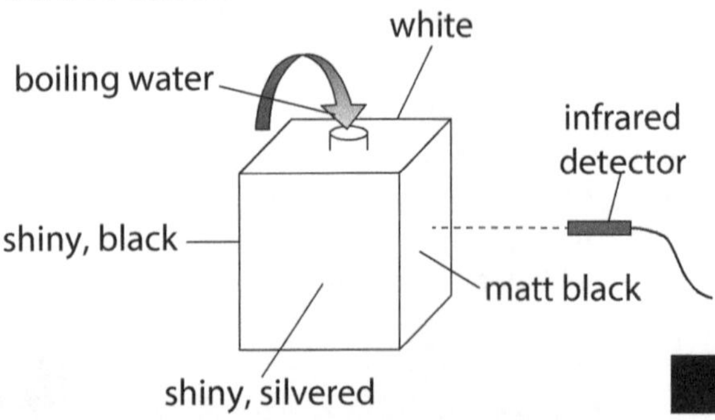

A46

Measuring energy, resistance, waves and force

Coll
Secure S

A detector used to measure the intensity of infrared radiation emitted by an object.

Tip: An infrared detector is used to measure the intensity of infrared radiation emitted by the different types of surface on a Leslie cube.

A4

Electrical circuits and electrolysis

Collins
Secure Science

Electrical components

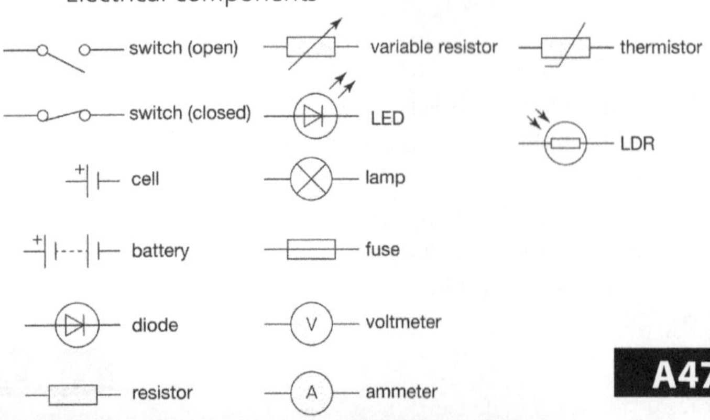

A47

Electrical circuits and electrolysis

Coll
Secure S

Components such as lamps, switches, resistors, diodes, thermistors and LDRs that use electricity in a circuit.

Tip: Electrical circuits must be complete for them to work. If the wire inside a bulb breaks then the circuit is no longer complete so the bulb will not work.

A4

Electrical circuits and electrolysis

Collins
Secure Science

Electrolysis

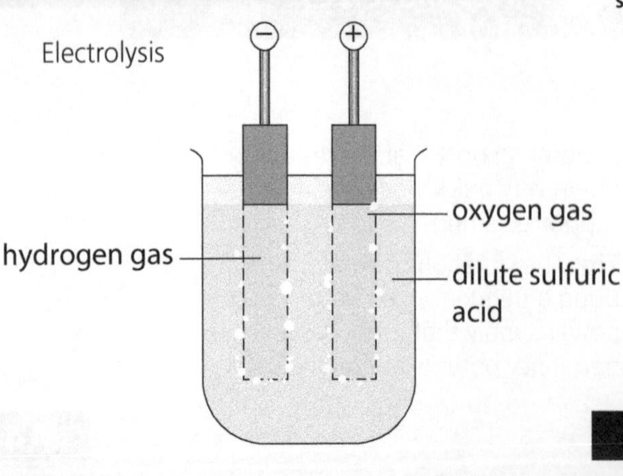

A48

Electrical circuits and electrolysis

Coll
Secure S

An electrical circuit through a solution or liquid containing charged particles that are free to move between electrodes as part of the circuit.

Tip: The electrodes used in electrolysis are sometimes made of graphite, which is not a metal but conducts electricity.

A4

Electrical circuits and electrolysis

Collins Secure Science

Joulemeter

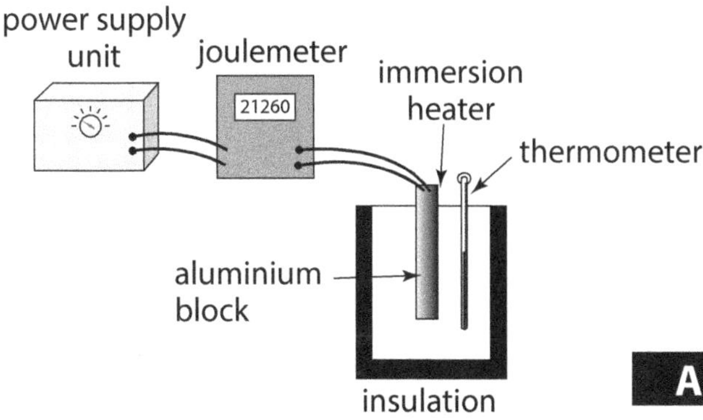

A49

Electrical circuits and electrolysis

Collins Secure Science

An electrical meter that measures the amount of electrical energy transferred to a component, such as an electrical heater.

Tip: Joulemeters are often used in place of an ammeter/voltmeter combination when determining the specific heat capacity of a metal block.

A49

Electrical circuits and electrolysis

Collins Secure Science

Electrical immersion element

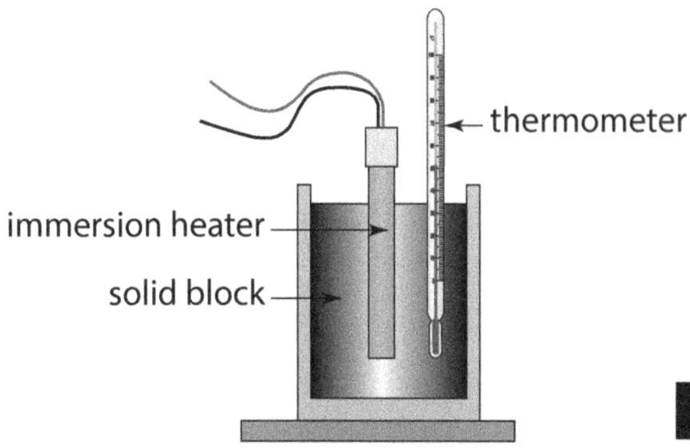

A50

Electrical circuits and electrolysis

Collins Secure Science

A 12 V electrical heater used to heat metal blocks or beakers of water.

Tip: The electrodes used in electrolysis are sometimes made of graphite, which is not a metal but conducts electricity.

A50

Indicators and tests

Collins Secure Science

Litmus paper

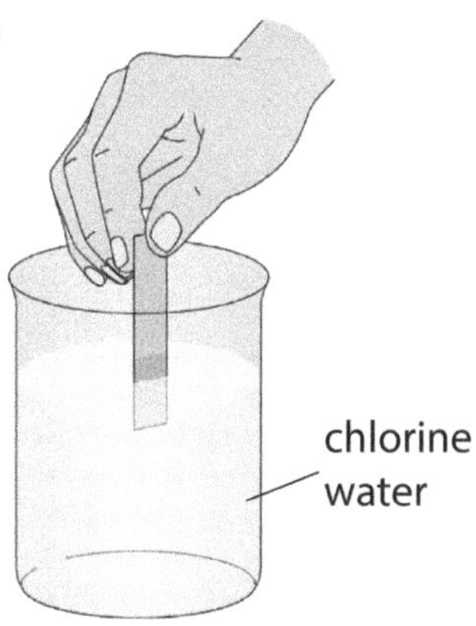

A51

Indicators and tests

Collins Secure Science

Litmus paper changes colour depending on the pH of the solution. Blue litmus paper turns red in an acid and red litmus paper turns blue in an alkaline solution.

Tip: Both colours of litmus paper stay the same colour when put into a neutral solution such as water.

A51

Indicators and tests

Collins Secure Science

Benedict's solution

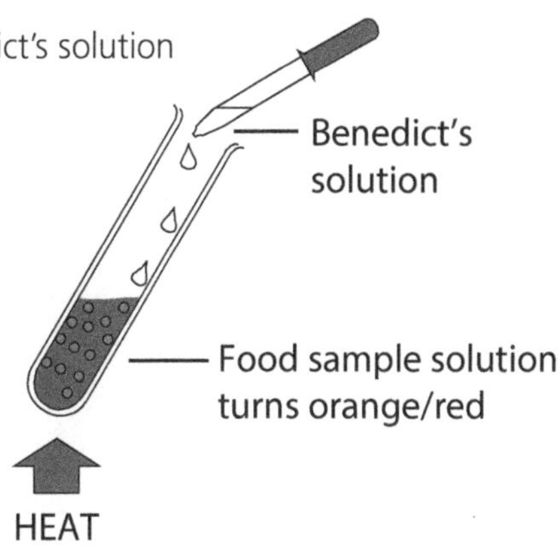

A52

Indicators and tests

Collins Secure Science

Test for sugar – blue Benedict's solution turns brick-red in a high concentration of sugar and yellow/green in a low concentration.

Tip: Benedict's solution changes from a clear blue colour to a green precipitate, then yellow, then orange-red, then brick-red, as the concentration of sugar increases.

A52

Indicators and tests

Collins
Secure Science

Biuret solution

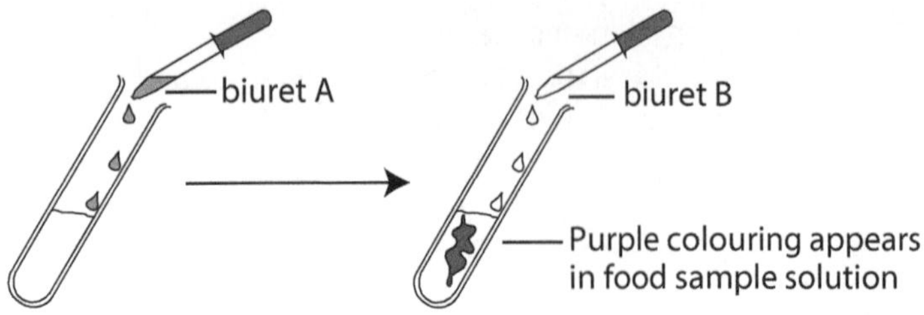

A53

Indicators and tests

Test for protein – blue biuret solution turns lilac when protein is present.

Tip: There is only one colour change with Biuret solution. It is a qualitative test.

Indicators and tests

Collins
Secure Science

Iodine

The brown iodine turns blue-black if starch is present in the food

A54

Indicators and tests

Test for starch – red-brown iodine solution turns blue-black when starch is present.

Tip: Iodine solution can be added to a food solution or dropped directly onto a solid food.

Indicators and tests

Collins
Secure Science

Sudan III

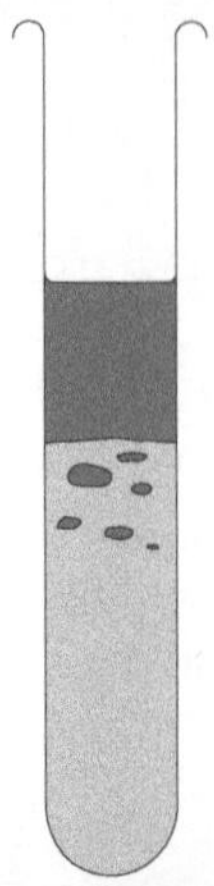

A55

Indicators and tests

Test for fat – a red stained layer appears above solution of the food if fat is present.

Tip: Sudan III needs to be in solution in order for this test to work.

Indicators and tests

Collins
Secure Science

Limewater

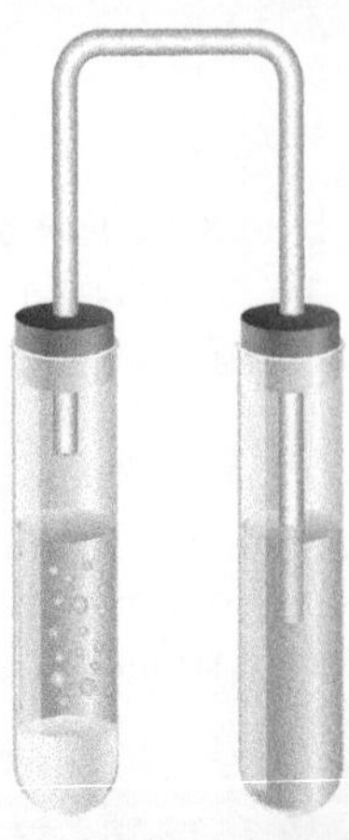

A56

Indicators and tests

Test for carbon dioxide – clear solution of limewater turns milky white when carbon dioxide is bubbled through it..

Tip: Always try to use fresh limewater as it absorbs carbon dioxide from the air and could give a false positive result.

Indicators and tests

Collins Secure Science

Lit splint

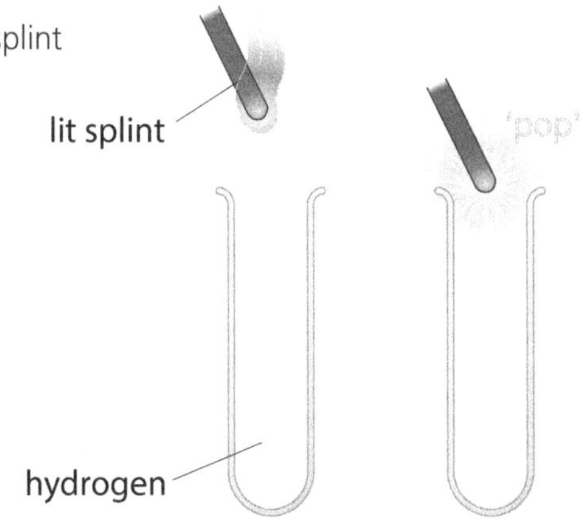

A57

Indicators and tests

Collins Secure Science

Test for hydrogen – hydrogen explodes with a squeaky pop.

Tip: Hydrogen gas should be trapped in a test tube by using your thumb to cover the end of the test tube. It is lighter than air and quickly escapes.

A57

Indicators and tests

Collins Secure Science

Glowing splint

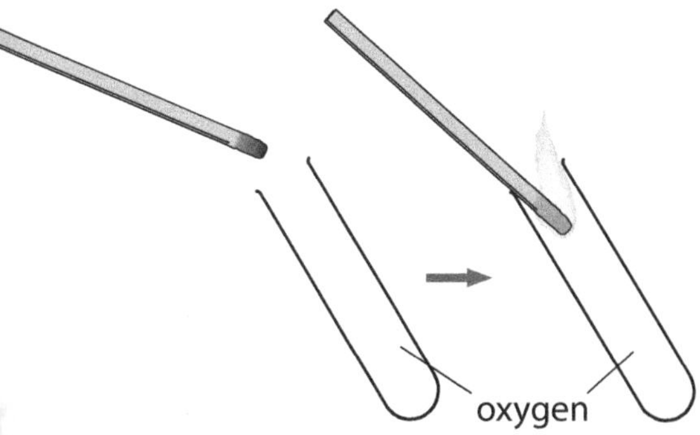

A58

Indicators and tests

Collins Secure Science

Test for oxygen – glowing splint will reignite if oxygen present.

Tip: A glowing splint is prepared by putting a fresh splint into a Bunsen flame, allowing it to catch fire, and then quickly blowing it out.

A58

Measuring mass and time

Collins Secure Science

Mass-stack

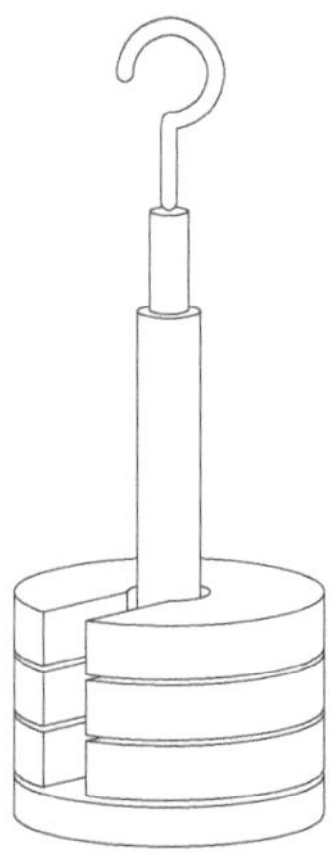

A59

Measuring mass and time

Collins Secure Science

Slotted masses arranged on a hooked holder/carrier.

Tip: A slotted mass-stack can be hung from a spring to investigate the stretching of springs and it can be used as a falling object to pull a trolley along the bench using string.

A59

Measuring mass and time

Collins Secure Science

Stopwatch

A60

Measuring mass and time

Collins Secure Science

Used to measure the time interval of an event, or the progress of a reaction or biological process.

Tip: Stopwatches can read to 0.01 s – but this is often unnecessary for experiments or events lasting many seconds or minutes.

A60

Measuring mass and time

Collins Secure Science

Top pan balance

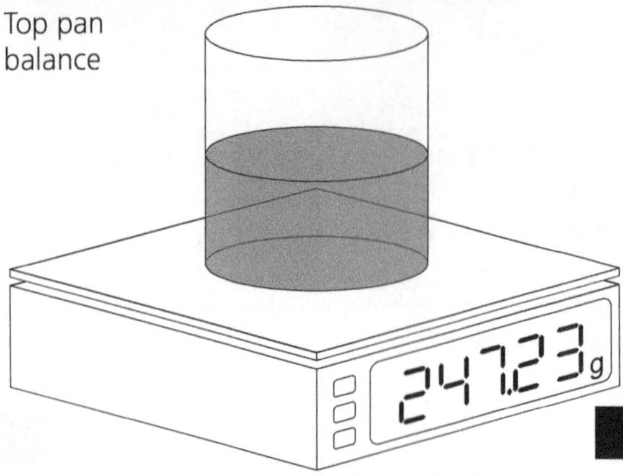

A61

Measuring mass and time

Used to measure the mass of solids and liquids in containers.

Tip: Uncertainty can be minimised by using a balance that is accurate to 2 decimal places. The mass of the container holding a liquid can be discounted using the zero button or by recording the mass of the empty container and taking it away from the total mass of the container and the substance.

Moving and holding substances

Collins Secure Science

Spatula

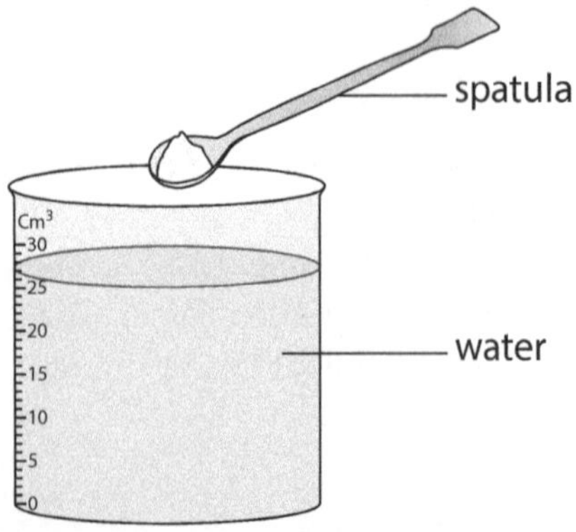

A62

Moving and holding substances

Metal scoop for roughly measuring small amounts of powdered substances.

Tip: Use a fresh spatula for each substance to avoid contamination of reactants. Do not use for stirring, particularly for hot substances.

Moving and holding substances

Collins Secure Science

Stirring rod

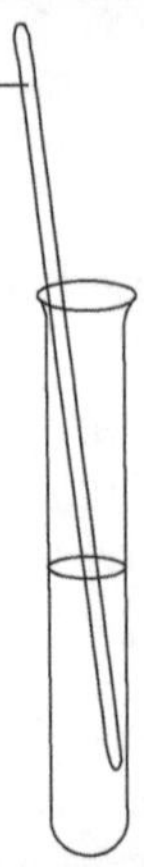

A63

Moving and holding substances

Glass tool for stirring liquids and powders in beakers.

Tip: Stirring rods are easily broken and prone to rolling off the bench – never leave on the bench alone. Use a different stirring rod for different solutions to prevent cross-contamination.

Moving and holding substances

Collins Secure Science

Clamp stand

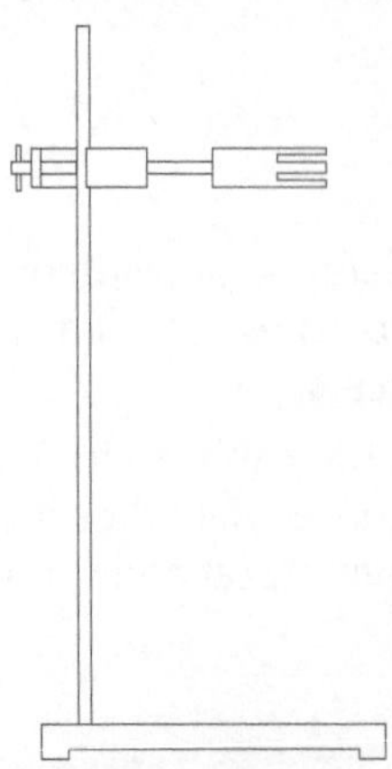

A64

Moving and holding substances

Metal tool that other apparatus can be attached to.

Tip: The boss connects the clamp to the stand, and the clamp holds tubes and other equipment.

Moving and holding substances

Collins Secure Science

Clamp and boss

A65

Moving and holding substances

Collins Secure Science

Metal tools to hold apparatus at a height above the bench.

Tip: The boss connects the clamp to the stand, and the clamp holds tubes and other equipment.

A65

Moving and holding substances

Collins Secure Science

Test tube/boiling tube holder

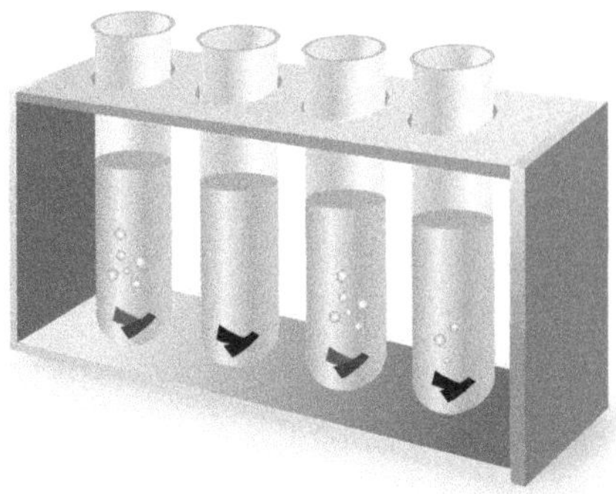

A66

Moving and holding substances

Collins Secure Science

Wooden or metal tool for holding tubes.

Tip: Take care when heating a tube: metal holders can get hot when heating and wooden ones can set on fire. Keep both well away from the flame of the Bunsen.

A66

Moving and holding substances

Collins Secure Science

Forceps

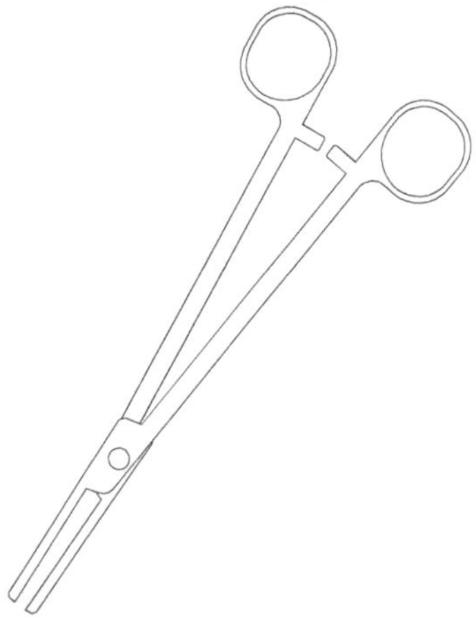

A67

Moving and holding substances

Collins Secure Science

Fine tool for holding tiny objects.

Tip: Forceps are often used to pull dead biological specimens apart.

A67

Moving and holding substances

Collins Secure Science

Tweezers

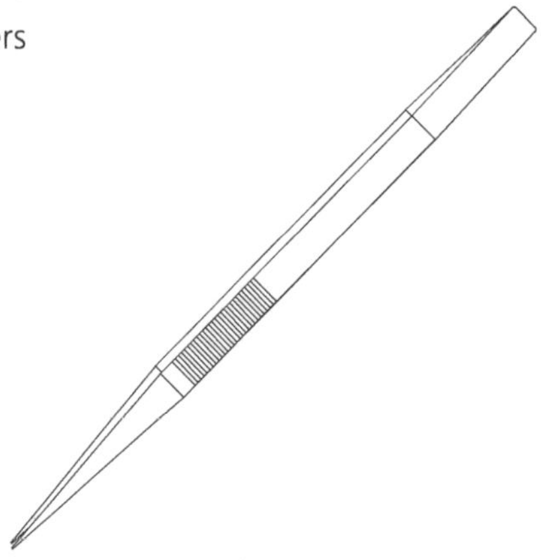

A68

Moving and holding substances

Collins Secure Science

Similar to forceps but can exert more force so can be used for separating tiny objects.

Tip: Fine tweezers are often used for manipulating very small objects, such as crystals, or holding indicator papers into gases or solutions.

A68

Moving and holding substances

Collins
Secure Science

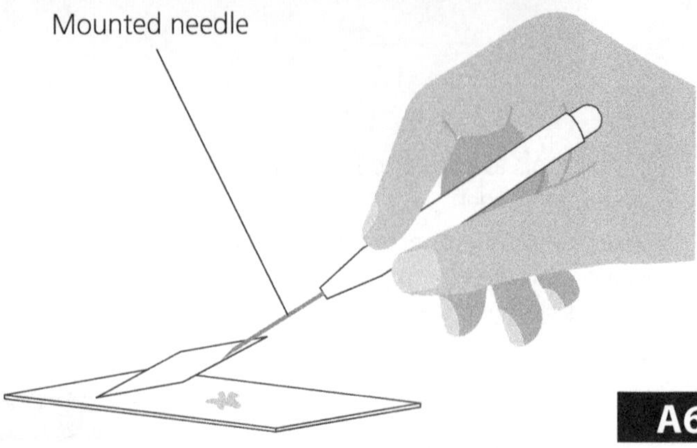

A69

Moving and holding substances

Coll
Secure S

Sharp length of metal set into a piece of wood. Used for moving tiny objects such as microscope coverslips.

Tip: Mounted needles are very sharp; handle with care.

A6

Moving and holding substances

Collins
Secure Science

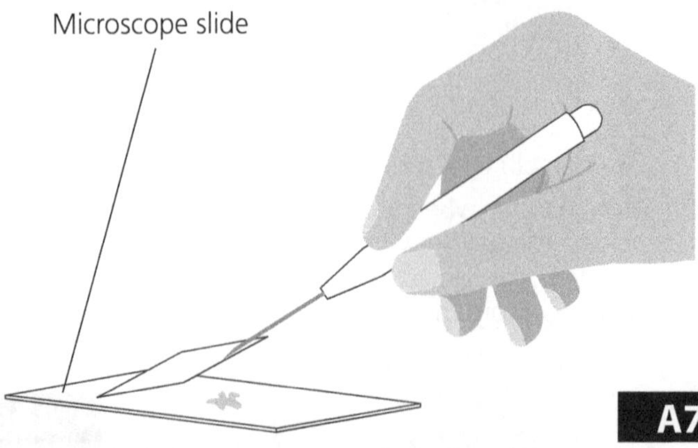

A70

Moving and holding substances

Coll
Secure S

Piece of glass that holds substances for viewing under a microscope.

Tip: Some microscope slides (cavity slides) have a shallow dip on one side and are used to observe microscopic organisms in pond water.

A7

Moving and holding substances

Collins
Secure Science

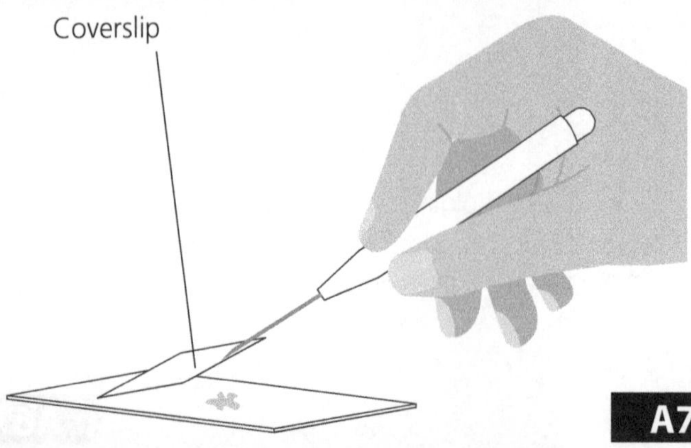

A71

Moving and holding substances

Coll
Secure S

Thin piece of glass that traps a sample on a microscope slide.

Tip: Coverslips are very fragile and crack easily; handle gently.

A7

Moving and holding substances

Collins
Secure Science

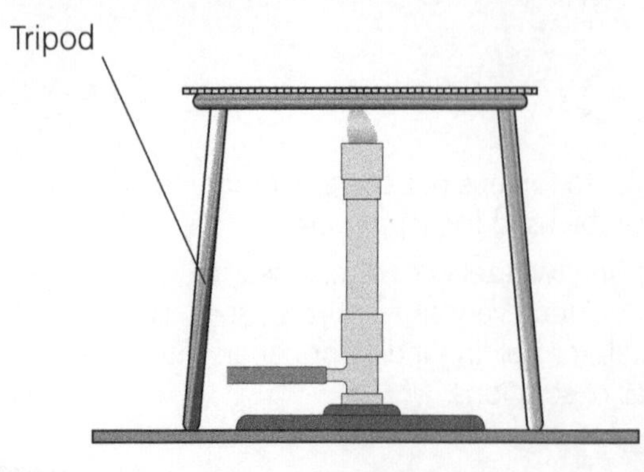

A72

Moving and holding substances

Coll
Secure S

Metal stand that holds containers above a Bunsen Burner.

Tip: Tripods get very hot during use and should be allowed to cool down before handling.

A7

Moving and holding substances

Collins
Secure Science

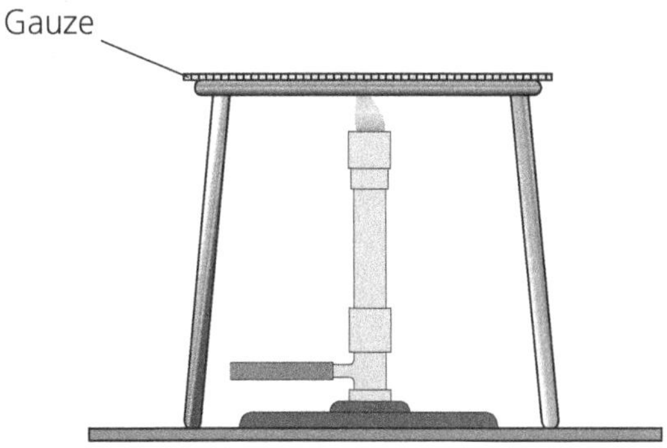

A73

Moving and holding substances

Collins
Secure Science

Sits on top of a tripod to support a container being heated above a Bunsen flame.

Tip: Gauzes get very hot during use and should be allowed to cool down before handling.

A73

Moving and holding substances

Collins
Secure Science

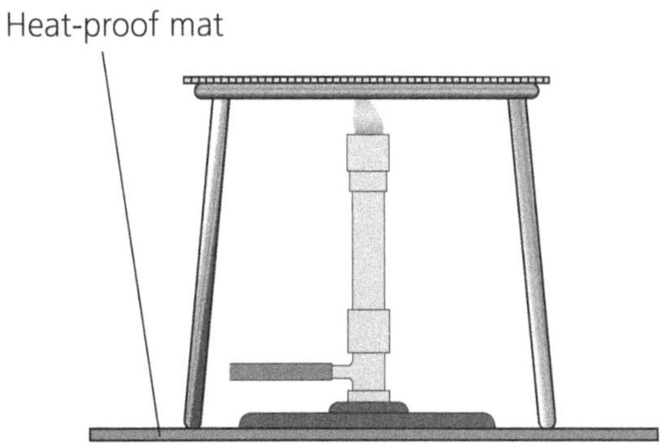

A74

Moving and holding substances

Collins
Secure Science

Sits underneath a Bunsen to protect the bench from heat.

Tip: Always use a heat-proof mat beneath a Bunsen burner.

A74

Moving and holding substances

Collins
Secure Science

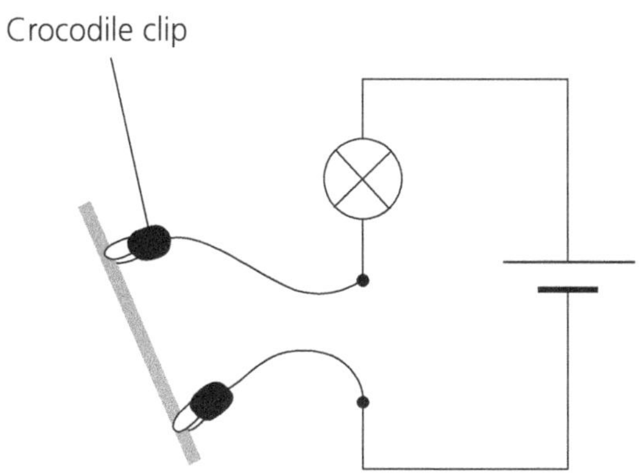

A75

Moving and holding substances

Collins
Secure Science

Used to attach metal components into an electrical circuit.

Tip: Some crocodile clips have an insulating cover over the end that is handled to prevent fingers from touching the metal.

A75

Moving and holding substances

Collins
Secure Science

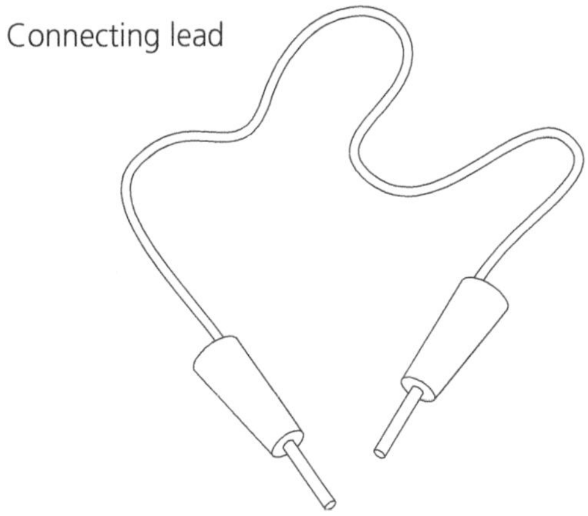

A76

Moving and holding substances

Collins
Secure Science

Insulated wire used to connect components in an electrical circuit.

Tip: The connectors at the end of connecting wires are easily pulled off the wire so they should be handled with care.

A76

Moving and holding substances

Collins
Secure Science

Teat pipette

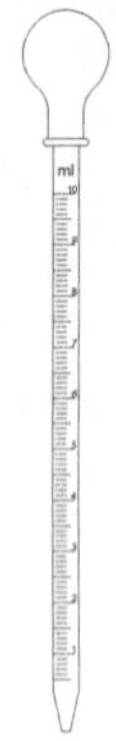

A77

Moving and holding substances

Col
Secure S

Used to transfer small quantities of liquid between pieces of glassware.

Tip: Make sure you have removed any bubbles from the liquid to get a more accurate approximate volume of liquid.

A7

Moving and holding substances

Collins
Secure Science

Tongs

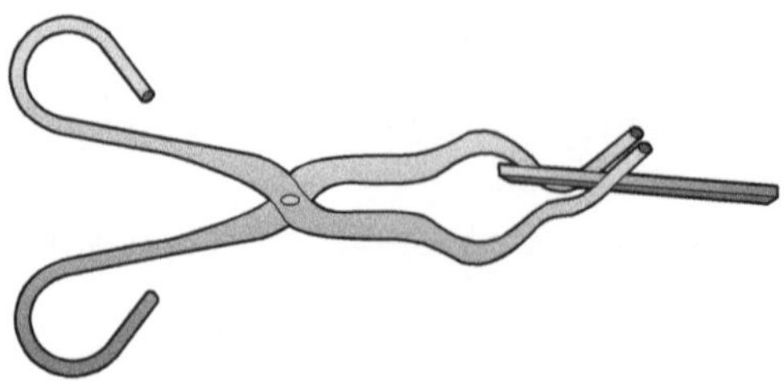

A78

Moving and holding substances

Col
Secure S

Tool used to grip and lift hot objects.

Tip: Do not hold tongs directly into a naked flame – they will get hot.

A7